Routledge Re

The Devil and the Vice in the English Dramatic Literature Before Shakespeare

The Devil and the Vice in the English Dramatic Literature Before Shakespeare

L. W. Cushman

Routledge
Taylor & Francis Group

First published in 1900 by Frank Cass & Co. Ltd.

This edition first published in 2018 by Routledge
2 Park Square, Milton Park, Abingdon, Oxon, OX14 4RN
and by Routledge
52 Vanderbilt Avenue, New York, NY 10017, USA

Routledge is an imprint of the Taylor & Francis Group, an informa business

© 1900 by Taylor & Francis

Publisher's Note
The publisher has gone to great lengths to ensure the quality of this reprint but points out that some imperfections in the original copies may be apparent.

Disclaimer
The publisher has made every effort to trace copyright holders and welcomes correspondence from those they have been unable to contact.

A Library of Congress record exists under ISBN:

ISBN 13: 978-0-367-14226-1 (hbk)
ISBN 13: 978-0-367-14227-8 (pbk)
ISBN 13: 978-0-429-03079-6 (ebk)

THE DEVIL AND THE VICE

IN THE

English Dramatic Literature

Before Shakespeare

THE DEVIL AND THE VICE

IN THE

English Dramatic Literature

Before Shakespeare

By

L. W. CUSHMAN

FRANK CASS & CO. LTD.
1970

Published by
FRANK CASS AND COMPANY LIMITED
67 Great Russell Street. London WC1

First edition	1900
New impression	1970

ISBN 0 7146 2055 6

Printed in Great Britain by Clarke, Doble & Brendon Ltd.
Plymouth and London

To my Father
and to the memory of my Mother

Preface.

An investigation of the devil and the Vice as dramatic figures in their historical relations has, so far as I know, never been made. Much has been written here and there on the subject, but all notions of the limitations as to when, where and how these figures appear or of the differentiation of their functions, are vague in the extreme. The prevailing opinion of the critics, historians and teachers of literature is quite uniform; it is, in substance, as follows: the devil enjoyed, both on the stage and elsewhere, a great and ever increasing popularity; the figure of the Vice was developed from that of the devil, or the Vice was simply the devil as buffoon and, as such, became the forerunner of the clown: he is also the forerunner of the villain, and of Punch.

A study of these figures at first hand has led to a new view of this subject; it is, in brief, as follows: the appearance of the devil in the non-dramatic as well as in the dramatic literature is limited to a definite range; as a dramatic figure the devil falls more and more into the back-ground, the Vice is distinct in origin and function from the devil and from the clown. It is not denied that these characters in the domain of the comical, on the one hand, and of egoism, on the other, encroach upon each other, but from this it does not follow that they are identical, or that the one is derived from the other. The devil, Vice, clown, fool and villain are parallel figures of quite independent origin and function.

The serious dramas of the period treated form the basis of this study. The writer has been fortunate in gaining access to the literature required, thanks to Professor Brandl for his recent „*Ergänzung*" to *Dodsley's Old Plays,* also for the permission to use his manuscript copies of some other, not yet reprinted, plays. Much of the material is, therefore, quite new.

In the quotations in the following pages, the orthography of the original has not always been retained, indeed, the accessible editions of many of the plays offer only a normalized text. Wherever, in the editions used, the lines are numbered, the references are accordingly to the lines, otherwise, to the pages only. The references to the three manuscript copies in the possession of Professor Brandl — *All for Money, Mary Magdalene,* and *The Tide tarrieth for no Man* — are in accordance with the paging of the old prints, thus: Ai, Aii, Aiii, etc., Bi, etc. This is not very satisfactory, but must suffice until these plays shall have been reprinted.

This study has been made under the encouragement of Professor Lorenz Morsbach of Göttingen, who has helped me not only with his advice but also in finding the widely scattered literature; he has, furthermore, authorized its publication in his *Studien zur englischen Philologie* and has kindly undertaken the correction of the proof-sheets. For these many favors, I take this occasion to express my warmest thanks.

My thanks are also due to my friend and enthusiastic co-worker, Mr. M. F. Libby of Toronto, Canada, for a first reading of the proofs, and to Professor Alois Brandl of Berlin, for the use of the proof-sheets of his *Quellen des weltlichen Dramas vor Shakespeare* some months before the book appeared, also for the use of the manuscript copies of several of the later Moralities.

Göttingen, July 19, 1899.

L. W. Cushman.

Contents.

Abbreviations of the titles of the plays.

A. V. = Appius and Virginia. (Dodsley IV.)
C. P. = The Castle of Perseverance.
Confl. = The Conflict of Conscience. (Dodsley VI.)
D. C. = The Disobedient Child. (Dodsley II.)
F. El. = The Four Elements. (Dodsley I.)
H. = Hickscorner. (Dodsley I.)
J. J. = Jack Juggler. (Dodsley II.)
K. C. = King Cambyses. (Dodsley IV.)
K. D. = King Darius. (Brandl. Quellen.)
K. J. = King John (Bale's). (Camden Soc.)
L. J. = Lusty Juventus. (Dodsley II.)
L. W. L. = Like Will to Like. (Dodsley III.)
Man. = Mankind. (Brandl. Quellen.)
M. W. Sc. = The Marriage of Wit and Science. (Dodsley II.)
Money or M. = All for Money.
M. M.[1] = Maria Magdalene (1580—90). (Digby Play.)
M. M.[2] = Maria Magdalene (1567), (by Wager).
Nat. = The Interlude of Nature. (Brandl. Quellen.)
N. W. = Nice Wanton. (Dodsley II.)
O. = Horestes. (Brandl. Quellen.)
Tide = The Tide tarrieth for no Man.
T. T. = The Trial of Treasure. (Dodsley III).
W. = Wisdom. (Digby Play.)
W. C. = The World and the Child. (Dodsley I.)
W. Sc. = The Moral Play of Wit and Science. (Sh. Soc.)
W. W. = The Marriage of Wit and Wisdom. (Sh. Soc.)
Y. = The Interlude of Youth. (Dodsley II.)

Part I. The Devil.

I. Introductory. The devil in the earlier, chiefly non-dramatic literature.

The devil as a dramatic figure presupposes a personal character having certain corporal attributes and having certain things to do; under the corporal attributes may be understood the outward appearance and name of the figure, under things to do, his actions and words. Appearance, action and words constitute the dramatic persona proper, consequently, references to the abstract principles of evil or to the dualism in nature, morals or religion, likewise, the mere mention of the devil as the agent of evil, but where he is not thought of as bodily present, are not taken into consideration; this study is concerned primarily with the representations of the devil as a figure on the stage.

The representations of the devil on the stage, it is reasonable to suppose, are largely traditional, hence the earlier non-dramatic literature may be regarded as the source of many details of the figure of the devil; from this great source the writers of the old plays probably derived many characteristic traits of the devil's figure, although, in this case, the possibility is by no means excluded, that the dramatic literature and the theological are but two quite independent streams flowing from a common source. That the Mystery-cycles, however, were not uninfluenced, especially by the *Cursor Mundi,* has been pointed out by Professor ten Brink. An examination, at the outset of this investigation, of the earlier non-dramatic English literature, and a collection of some typical examples will help, therefore, to determine, in a measure, what the prevailing notions in the Middle Ages concerning the figure in question were.

Sources. — The non-dramatic literature, in which the devil is especially to be found, is almost entirely theological: Homilies, Legends of the Saints, biblical Histories and didactic treatises. In the matter of introducing the figures of devils, the writers in the Middle Ages were not as free as is sometimes believed. In the Legends, as elsewhere, tradition plays an important part, thus checking a tendency, inherent in the very nature of these stories, to become with time more and more extravagant. As a rule, the appearance of the devil is confined to certain Legends; in many, the figure is wanting in all the versions.

In the Homilies, Histories and didactic tracts the observation is here to be recorded; first, that devils appear with remarkable regularity only in certain great scenes: the *Fall of Lucifer,* the *Fall of Adam and Eve,* the *Temptation of Christ,* the *Harrowing of Hell* and *Doomsday*; second, that they do not appear, except in rare instances, in certain other great scenes: The *Murder of Abel,* the *Flood,* in the story of Pharaoh, of Herod and the like. Furthermore, in the stories of the childhood and education of the Virgin Mary and of the childhood of Jesus, the devil plays no conspicuous part. Thus it is evident that the devil is a traditional figure, and that the ultimate source of this tradition is the Bible, including the Apocrypha, as understood and interpreted by the early Church Fathers. This observation holds good also, as will be seen later, in the Mysteries. In the mediaeval religious literature, the dramatic as well as the non-dramatic, with the exception of the Legends and of the later Miracles (see below), the occurrence of the figure of the devil is confined to biblical precedents.

The treatment of the devil in the abstract. — It is customary, particulary in didactic literature, to refer to the devil in a general way, as the source or principle of evil. Expressions of general reference are exemplified in phrases, such as: „deofles wise", *Blickling Hom.,* p. 55, „fram deofles anweald", *Aelfric,* I, 120, „þurhh deofell", *Orrm,* 11416, etc. etc. Physical evil, especially if mysterious, is also ascribable to the devil; thus Richard Rolle of Hampole explains in *De Natura Apis* how honey in the hives breeds worms: „þe deuile turnes it to wormes", 25.

In *Beowulf* the identification of physical evil with a spirit, like Grendel, as well as the propitiation of the „gast bona", 177, and the assigning of Grendel finally to the company of fiends: „deofla ȝedræg", 757, „on feonda ȝeweald", 809, were undoubtedly in the heathen original; the explanation, that the worship of the „gast bona" is devil-worship, and that evil spirits and monsters are the offspring of Cain, is due to the clerical revisers of the poem. Grendel is especially designated as „feond on helle", 101, as „ellor gast", 808, as „goddes ansaca", 1683, and as a descendant of Cain:

   ... „þanon onwoc fela
ȝeosceaft gasta: wæs þæra Grendel sum", 1267.

This is an interesting instance of an early contact of heathen and Christian demonology.

A satirical use is made of the devil's name by Aelfric in his sermon on the *Assumptio Sce. Mariae Virginis;* idle words, anger, songs, stories, and the like, are called the „devil's seed", II, 163. In his sermon on the *Prophet Jeremiah* still stronger expressions are used: „þe ȝeolewe claþ is þe deofles helfter" (halter), foolish women are called the „devil's mouse-trap" and „blanchet" (white powder) is the „devil's soap", I, 53.

Abstractions, however, may be personified, and of personifications dramatic characters may be made. The World, the Flesh and the Devil, and the Seven Deadly Sins are often treated in the old clerical writings. The World, the Flesh and the Devil are called our three foes: cf. *þe wohunge of ure Lauerd*, p. 277, the Old English Homily, *Induite vos armatura die*, I, 243, *Cursor Mundi* 10104, 23746, 25730, etc. etc. The Seven Deadly Sins are referred, in their origin, to the seven spirits which were cast out of Mary Magdalene: „þe seven full gostes þat ich nu embe was, waren þe difles giltes, þat ure drihten drof ut of seint Marie Magdaleine", Old English Homily, *In media quadrogessima*, II, 87. A common classification of these powers is the following, as given by Wyclif:

„Pride, envy and wrath ben synnes of þo fende; wrath, slauth and avarice ben synnes of þo world, avarice and glottenye and þo synne of lechorye ben synnes of þo flesche“, Works, III, 119. cf. Chaucer, *The Tale of Melibeus,* 2610 „the three enemies of mankind, that is to seyn the flessh, the feend and the world“. The personification is complete in Langland's *Vision of the Seven Deadly Sins*; they are all masculine, excepting Pride, and each is described by some characteristic trait. In this particular instance, they are much affected by the sermon of Reason and repent their sins: Superbia will don a hair shirt, Luxuria will „drink but with the duck“, Invidia confesses to „backbiting“, Ira, that he has incited many to quarrel; Avaritia thought that „rifling were restitution“, not understanding French, Gula promises „to fast“ and to eat no fish on Friday without leave of his aunt Abstinence, whom he has hitherto cordially hated, Accidia is a priest who „can rhymes of Robin Hood“ but not of our Lord or Lady.

Personifications like these, together with the real personal devil, figure prominently in the Moralities, but in the Mysteries no bad powers are personified, excepting Mors. The Coventry Plays have a number of figures of good powers like Pax, Contemplacio, etc., Chester has Thrones, Virtutes etc., Townley, Trinitas.

The treatment of the devil concretely. — The biblical Histories and the narrative portions in the Homilies, being more or less of an epic character, approach much nearer to the dramatic. The devil is treated here, as also in the Legends, as a real person in active relations to other persons and is thus made capable of furnishing motifs for dramatic action.

The Fall of Lucifer is always thought of realistically; the change in the appearance and name of the fallen angel is especially definite. This scene, with attendant circumstances, is fully described in the *Cursor Mundi*; Lucifer is represented speaking, he declares his intention to place his throne on the north side of heaven over against the seat of Most High; God shall have no more service from him. It is then related that Michael arose and fought against Lucifer and cast him out of heaven:

„þis is þe feind þat formast fell,
þoru his ouergart (pride) in to hell,
Fra þan his nam changed was
For now es he cald Sathanas". 477—480.

(The Göttingen Ms. has „foule Sathanas", the Trinity, „From Lucifer to Sathanas"). *The Genesis and Exodus* likewise relates that Ligber, i. e. Lightbearer, took his flight and set his seat on the north side of heaven between heaven and hell and became, in consequence, a black drake:

„ðo wurð he drake ðat ear was knigt,
ðo wurð he mirc ðat ear was ligt". 283—4.

The blackness of the devil is often referred to: „þe blake deofol", *Soules Wearde*, 251, „swarttore þene euere ani blowȝman". *Early South English Legends*, I, 245/165; likewise, his loathsomness: „þe laþe Sathanas and Belzebub þe elde", *Poema Morale*, 285, „þe loþe gast", *Orrm*, 11355.

In the Temptation of our first parents the form of the devil, here generally Satan, is already determined by the biblical account, namely, that of the serpent. Satan, seeing Adam in bliss is filled with envy because man has been created to take the place in heaven that he himself has lost. In the *Genesis and Exodus* it is thus tersely related that Satan „wente into a wirme and told Eve a tale", 321. The serpent here is not the devil himself but an assumed form; thus according to *Cursor Mundi*:

„þat wili warlau him heild on drei
And ganid noȝt cum him to nei,
Nameli in his auen schap . . .
for-þi a messager he send . . .
To þis he ches a littel best,
þe quilk es noȝt vnwiliest,
þe nedder þat es of a scaft,
þat mast can bath on crok and craft . . .
þis nedder forth þat he ne blan
Bot in hijs slught (skin) was self Satan", 731—745.

The Temptation of Christ is in all accounts told with utmost fidelity to the biblical story. The only liberties the writers allow themselves being the use of certain epithets;

Aelfric designates the tempter as bloodthirsty (wælræw), I, 192, the *Cursor Mundi,* as „þe warlau wili", 12930; The outward appearance of Satan in this scene is not described.

The story of Job afforded, ready at hand, a good devil-scene, but this plot seems not to have attracted much attention. *Aelfric,* however, gives a full account of the story in almost the exact words of the Bible. He designates the devil as „manfulla" and as the „ealda", who goes from the presence of God and destroyes in one day all of Job's possessions, II, 450.

Judas. — The devil is spoken of as actually entering Judas, as is recorded in the *New Testament;* thus in *Aelfric:* „Huæt se deofol into Judan bestap", II, 242. According to the *Cursor Mundi* the Savior gave Judas a morsel of bread and with that morsel „crep in him Sathanas", 15388, and after Judas had hanged himself, the fiend hurled him into hell, 16528. The carrying off of the souls of evil doers to hell, as in the case of Judas, is a function of the devil especially developed in the Mysteries and in some of the Legends.

With other persons, as has already been intimated, the devil has but little to do. Cain was according to the scribes of the *Beowulf,* the ancestor of the evil monsters, according to the *Cursor Mundi* he was the „devil's food", but, strange to relate, the temptation of Cain to kill his brother was without the special instigation of the devil, 1056. Pharaoh appears to be regarded by *Aelfric* as an incarnate devil: „Pharao getacnode þone oywan deofol", II, 200.

The *Doomsday* does not possess the interest that the other great devil-scenes do, it is not so often described and, on the whole, lacks in definiteness. The devils take no very active part. The accusers of the guilty ones on that great day shall be God, our conscience, the world and the devil, *Cursor Mundi,* 26711; but accusers are hardly necessary, as the wicked, especially suicides, shall rise with a maimed body, while the righteous shall rise whole and without blemish, *Cursor Mundi,* 22836. As a matter of fact, Doomsday is the end of all things, for the devils as well as for men; thus in *Cursor Mundi* the signs of Doomsday are described, the eleventh sign being a rainbow which shall „dump the deuls þider in", that is, into hell, 22643.

The Harrowing of Hell. — The discussion of this subject
has been reserved until the last of this series of devil-scenes,
partly as a matter of convenience, but chiefly because of its
importance. The descent of Christ into hell, the release of
the souls and the binding of Satan, are treated with unusual
fullness and detail, and had undoubtedly taken a strong hold
on the mediaeval imagination. *The Descensus Christi ad Inferos* is a work dating probably from the third century, it is
also contained in the apocryphal *Gospel of Nichodemus.* The
basis of the tradition appears to be the *24th Psalm*; the verses
7—10 being cast in an antiphonal form, possess of themselves
a sort of dramatic interest: „Lift up your heads ye everlasting
doors and the king of glory shall come in. Who is the king
of glory?" etc. These words, according to the old patristic
exegesis, were made to refer to the journey of Christ into
hell; Christ comes with the victorious banner of the cross in
his hand and utters his challenge; hell demands who he is and
what he intends. This is the principal scene about which
cluster several subordinate ones: the joy of patriarchs in limbo,
the consternation of the devils, the council of the devils,
contest between Christ and Satan, the final binding of
Satan by the Savior or by the Archangel Michael.

The *Dominica Pascha,* one of the Blickling Homilies,
A. D. 971, describes the scene with minute detail, and is, in
part, in dialogue-form. The spirits in great alarm ask:
„Hwanon is þes þus strang & þus beorht & þus egesfull?"
Turning to Satan they ask, „gehyrstu ure aldor? þis is se
ilca þe þu longe for his deaþe plegodest ... ac hwæt wilt þu
nu don?" They tell Satan that, in compassing the crucifixion
of Christ, he has overreached himself: „tohwon læddest þu
þeosne freone & unscyldigne hider?" Thereupon was heard
the lamentation of the hosts of hell („seo arelease helwarena
stefn"), the iron bolts of the locks of hell were broken, and
Christ felled the old fiend without delay and threw him bound
into the abyss.

Aelfric treats this subject briefly, giving, as a reason
for Christ's harrowing hell, the fact that the devil had, in
instigating the crucifixion, forfeited all claim to the souls
in limbo; like a greedy fish he had snapped at a baited

hook. The devil did not understand that Christ was divine, I, 216.

The account given in the *Cursor Mundi*, 17980, seq., is both full and vigorous. At the first approach of Christ the usual contention between Satan and Hell (here personified) takes place; Satan orders Hell to make ready to receive Him that boasted himself to be the son of God. Hell demurs and demands of Satan, who is here both prince and porter, why he would bring Christ into hell. Satan explains that he had caused the Iews to crucify Him and mingle vinegar with gall. Hell is afraid and forbids Satan to bring Christ in, maintains that Satan knew not what he was about and tells him that he is but a „faint fighter". Hell casts Satan out and bolts the doors, Christ without repeats angrily his demands, Hell seizes and sharply snubs Satan:

> „Hell hint ere þam þat Gerard grim
> And selcut snarpli snibbed[1]) him."

And thus while Sir Satan and Hell made their „murnand man" (contention), the King of Bliss had his will.

It will be appropriate to consider in this connection the poem entitled the *Harrowing of Hell*, A. D., 1290, (ed. by E. Mall, 1871, and by K. Böddeker, 1878). This poem is a debate or contention between Christ and Satan and, in all probability, is the earliest devil-scene in dramatic form. That it was intended to be acted is probable; it is not only similar to the corresponding play in the great Mysteries, particularly Townley and Coventry: cf. Mall, but furthermore, the *Christi descensus ad Inferos*, acted before Henry VII at Winchester in 1485, was, as Böddeker remarks, probably this very piece.

The poem is in dialogue form and contains eight persons: Christ, Satan, the Janitor (a devil), Adam, Eve, Moses, David and John. Satan disputes with Christ; being beaten in argument, he complains of his misfortunes:

> „Ich haue had so michel wo,
> That I ne recche, whider I go."   119—120,

---

[1]) To snub = to scold, blame, can also mean to tie short, e. g. „snubbing post".

and threatens, in order to make good his losses, to „go from
man to man". This, however, Christ will prevent: „So faste
shall I binde þe", 127; only „þe smale fendes, þat ben un-
stronge", 131, shall be permitted to go among men to tempt
them. The actual binding of Satan does not here take place,
nor does the angel Michael appear.

The little devils, „þe smale fendes" could easily and with
good effect be represented on the stage; that they were
actually present, however, is not indicated in the poem as we
have it, nor do they appear as such, nor are they even
mentioned in the *Harrowing of Hell* of the Mysteries.

Another interesting figure in this poem is the devil-
porter of hell. His chief trait is cowardice; he leaves the
gates to be guarded by whoever will and, with certain side-
remarks, runs away:

> „Ne dar I her no lengor stonde;
> Kepe þe ʒates, whoso mai,
> I lete hem stonde and renne awei." 140—2.

This figure belongs to a well-known type; cf., for example,
the Porter in the *Conspiracy to take Jesus,* York Plays, 226,
and in *Macbeth,* II, 3. A devil as the porter of hell is mentioned in a humorous way in the Townley Plays, 379/373, seq.,
otherwise, this rôle is not played by a devil.

Forms. — The forms in which the devil makes his
appearance are, on the whole, not greatly varied nor definitely
described. References to the devil as a serpent or to the
devil in a serpent's form are common. Thus the sermon,
Estote fortes in Bello, explains the resemblance of the devil
to the serpent: „diabolus nominatur hic serpens propter tria
invidia; tabescit sine strepitu, serpit, quod pungit veneno afficit",
Hom., p. 153: cf. also *Aelfric,* I, 16. Other forms of the devil
are sometimes mentioned in the Homilies, but these expressions
are, for the most part, figures of speech. According to the
sermon on *Psalm CXIX,* he is a hunter, while men are the
wild animals and the world is the wilderness, *Hom.,* II, 209.
In the *Dies natalis Domini,* it is taught that our foe the devil
assumes a variety of forms (geres), sometimes that of a fox,
sometimes that of a wolf, *Hom.,* II, 35. *Aelfric* mentions an

invisible devil (ungesewenlic), II, 176 and 454. Invisibility is
a motif occurring often in the Legends and in the Morality,
Mankind.

*The Legends of the Saints and Homilies on particular
Saints.* — The legendary literature differs from that already
considered in that the devil-scenes are not based upon biblical
sources; in the Legends, devils may be introduced anywhere,
but even here, as has already been pointed out, devils are
not introduced at random. The nature of the contrast between
devil and saint and the reason for the devil's bitter hatred
are discussed at length by Roskoff in his *History of the
Devil*; the whole matter may be summed up briefly thus: the
saint possesses divine virtues and practises asceticism. By the
former, the saint, though human, acquires superiority over
the devil; the devil is, in consequence, peculiarly aggravated;
by the latter, the saint is made peculiarly liable to the
attacks of the enemy. The devil of the Legends is the
embodiment of the sins that so teased and tormented the
holy devotees.

The cross is likewise an object of the devil's hatred.
The earliest representation of this in English is in Cynewulf's
Elene 900—934. The „lacende feond" comes flying in the
air, he bewails the finding of the cross, he recounts bitterly
how Christ had already diminished his realm and bereft him
of his property and threatens to instigate „another King"
against the faith. Judas rebukes him and he vanishes.
Cynewulf found this description, for the most part, already
in his Latin original. The *Cursor Mundi* has the same story:

> „Quen funden was þis hali croice,
> þe warlau said wit voice,
> A ha! Judas quat has þou don?", 21813—5.

The forms in which the evil one was accustomed to show
himself are various. In most cases, and always, unless there
is some reason for the contrary, he appears in his own proper
shape, but, unfortunately, details regarding his form are not
very abundant. When he appears in his own proper shape,
he is described simply as the „devil", „the black devil",

„loathsome devil", „foul fiend", and the like. Stench is often mentioned as an accompaniment of the devil: cf. *Early South. Engl. Legends,* 208/264, 283, 209/301, 312: cf. also the description of the enraged dragon: „stonc þa æfter stane", *Beowulf,* 2289, this dragon also spouted fire: „gledum spiwan", *Beowulf,* 2313.

Assumed forms or disguises are frequent, but, as in the case of the real form, they are not explicitly described. The disguises are assumed either for the purpose of tempting the saint or of frightening him. To St. Dunstan the devil appeared in the form of a fair woman, to St. Andrew, as a fair damsel to be shriven, to St. Theodora, in a wild beast's likeness, to St. Martin as a „cruenda bestia"; St. Madwen saw a devil once in the form of a little black boy clinging to Bishop Cheuin's foot.

Occasionally the devil is described as a dragon: „ænne dracca", *Aelfric,* II, 176, *Hom.,* p. 251, „dragoun", the *Legend of St. Margarete,* as a dragon spouting fire in *St. Magarete* and in *St. Bartholomew*: Fuyrie speldene al stinkende out of his mouth he blaste and fuyr of brumston at his nose". It is worthy of note, by the way, that the Legends, Homilies and Bible stories give no hint of the devil having horns, a tail or a cloven hoof; such details are left for development to the pictorial and histrionic arts.

The most elaborate description of a devil's form is given by Osbern Bokenam in his *Life of St. Margarete,* 1443; St. Margarete in prison prays that she might see her enemy; thus he appears:

> „A huge dragoun, glastering as glas,
> Sodeynly from a corner dede apere
> Of the presoun, with an horryble chere;
> His hairs were gylt, his beard was long,
> His teeth of iron were mighty & strong;
> Out of his nosethrylles foul smoke he blew,
> His eyen glastryd as sterrys by night,
> His tongue over his crowne he threw,
> In his clawys a swerd burnished bright,
> And anoon the presoun wex full of light

Of the fire wych out dede renne
From his mouth & fast gan brenne." 450—61.

This description is, however, not new. Bokenam here follows his Latin original: cf. Horstmann, *Bokenam's Legenden*, IX. The versions of this Legend in the Ashmolean MS. and in the Auchinleck MS. contain a similar, but not so full, description of the dragon.

The saint was not without reason frightened at this monster, particularly so, as he proceeded to swallow her, but, unluckily for him, he swallowed also her cross with her; he, therefore, burst in two and she „escaped harmless" and „thus had the victory". That this monster was a devil, there is no doubt, his name was Ruffyn. „Another deuyll", brother to this one, appears and explains how Ruffyn had assumed the dragon's form: cf. 483. In the Ashmolean version the name is Geffron.

The devil always defeated. — The saints often find them-selves in fearful straits in these onslaughts of the evil one, but they are always able, with infallible certainty, to deliver themselves out of all dangers; the sign of the cross, a prayer, a relic, the breathing of Christ's name or merely thinking thereon at the right moment, has the effect of silencing the devil or of putting him to flight. Even the accidental swal-lowing of a cross caused the body of the monster, according to the *Legend of St. Margarete,* to burst open; the saint escaped unharmed, the monster died. Thus according to the prevailing idea of the Legends the devil is doomed to an everlasting, though ineffectual struggle. Sometimes he com-plains bitterly of his ill luck, as in the *Elene,* sometimes he is forced to describe his evil works and ways, as Satan does in the *Legend of St. Margarete* or the devil in the *Legend of St. Dominic,* sometimes he is forced, instead of doing the harm he intended, to serve the saint, as in the *Legend of St. James.*

Often, and always to his great disadvantage, the saints or Christ encounter the devil with physical force. In the *Life of St. Catherine* it is related how Christ overcame the fiend and shaved his head, „To scafet his heaued, E. E. T., 56/1190.

At one time there came a flock of devils through the air to carry off St. James to their master, Hermogenes; the saint, however, simply ordered angels to bind the devils in hot chains. The most interesting example of this sort is the popular story of St. Dunstan, who seized the devil one morning by the nose with a pair of hot tongs.

With sinners, on the other hand, the devil has usually more success. *Aelfric* relates how a furious devil seized Egeas, the persecutor of St. Andrew, and threw him to the earth in the sight of all who were present, I, 598; he also relates how an importunate monk once got the consent of St. Benedict to leave his cell; the monk was no sooner outside the monastery than he began to cry for help, for he saw a dragon. *Aelfric* adds: „forðon þæt wæs se ungesewenlica deofol“, II, 176. Similarly in the *Legend of St. Brendan,* the devil captures and carries off a bad monk, in the Legend of St. Agnes he strangles the son of the Prefect, and is thus instrumental in rescuing the saint. *The Legend of St. Edmund* relates how the saint one day saw a flock of devils like crows in the air, gaily tossing about „a luyte blac sak“. In this sack, according to the statement of the saint, was the soul of a man from Chalgrove who had just died and whose body was still lying warm on his death-bed.

As a matter of course, the devil lays claim to the souls of all bad men. *Aelfric* relates that St. Drihthelm, a pious Northumbrian, was once conducted to hell and there saw how the devils were engaged in bringing in souls: „sum þæra wæs preost, sum læwede mann, sum wimman and ða deoflu sægdon hlude hlihhende þæt hi þa sawla for heora synnum habban moston“, II, 350. But, as a general thing in the Legends, the devil is cheated of his booty. In the *Legend of St. James* it is related that the devil once induced a young man, a protégé of this saint, to mutilate himself. The young man afterwards commits suicide, and, consequently, must go to hell. The devil comes to claim his property, but the saint makes a counter-claim, „þu berst more þane þin owe“. There follows a bitter contention between the two over the soul in question. The devil, apparently, is about to make good his claim, but is suddenly baffled by a miracle, the young man is brought to life again, *Early South Engl. Legends*, p. 44. Chaucer, in the *Friar's Tale*, gives a description of a devil dressed as a yeoman and riding through the country. The Friar and the Sompnour, being rivals and enemies, seek to demolish each other; the Friar relates how the devil once got possession of a Sompnour. The Sompnour in great rage retaliates; he tells in his *Prologue* a filthy anecdote and describes, where the Friars have their „nest" in hell. Satan is here represented as having

> ... „a tail
> Brodder than of a carrick is the sail".

Dialogue: Cries. — The Legends contain, occasionally, passages in form of dialogue between saint and devil. The best example of dialogue is in the *Life of St. Dominic*, A. D. 1290. Here the names of the speakers, „Diabolus", „Dominicus", are given, *Early S. Engl. Legends*, p. 285, 286. *The Life of St. James, Early S. Engl. Legends*, p. 35 and the *Legend of St. Serf and the Devil* in Andrew Wyntoun's *Chronicle* contain likewise good examples of dialogue.

Crying and shouting appears not to be a characteristic of the devil in the Legends, except in cases where he has been seized with hot tongs or bound in hot chains; in this case his howling is justifiable.

Intrigue. — In his attempts to circumvent the saints the devil sometimes resorts to intrigue. For example, the devil hates especially a certain St. Madwen. He goes to a certain Bishop Cheuin, who had been doing penance as a hermit for seven long years and relates to him that St. Madwen had converted a band of thieves promising them immediate entrance, that is, without penance, into heaven; „You see", said the fiend, „your long penance is altogether useless". This infection works like poison, the Bishop collects a following and marches out to punish the saint. As he approaches, the saint sees the devil himself in the form of a small black boy, hanging to the Bishop's left foot and whispering evil suggestions into his left ear. The saint simply prays to God for help, the Bishop immediately sees his mistake and returns humbled and instructed to his cave, *Lives of Engl. women Saints*, p. 93, 94.

In this connection may be mentioned the occurrence of two examples of the Faust-motif. The first is the well-known story of *St. Theophilus,* who through the agency of a Iew, makes a contract with the devil that, for great riches, he will deny Christ and the Virgin Mary. A second example is given in Aelfric's „*Life of St. Basileus*". The devil instigates a young man to love a maiden who has been consecrated to the service of God; he will help the young man only on condition of receiving a written contract: „ac wryt me nu sylf wylles þæt þu wiðsaca Criste", p. 379. He then fires the maiden with love for this young man. Later the saint undertakes to free the young man from the power of the devil and to take him again into the church. The devil, in the mean time, attempts to tear the young man away from the saint crying, „He is mine! I have his written agreement, (his hand-ge-wryt)". „Very well", says the saint, „we will all shout, Kyrie eleison". The result is, the paper falls out of the air into the hands of the saint.

The dramatic value of the Legends. — According to the above cited examples, it will be easily seen that the Legends are almost entirely wanting in dramatic material. First, the necessary conditions of dramatic interest are lacking; the Legends contain, it is true, great contrasts, action and passion, but a situation of suspense is in every instance rendered impossible by the absolute holiness of the saint. In the face of such infallibility what can a poor devil do! Second, real humor is lacking; many scenes, indeed, in the Legends are, according to modern standards, comical or grotesque, but that they were originally intended to be so, particularly in the earlier forms, is doubtful. For such scenes to become regarded as comical, repeated exhibition before the public would be necessary; this was not the case with the Legends.

That the devil-scenes in the Legends have had so little influence upon the English drama, apart from the few drama- tized Legends, or Miracle Plays, is owing partly to the nature of the subject-matter of the Legends, and partly perhaps to the prevailing taste in different districts of England. As Ten Brink remarks, *History of Engl. Lit.,* II, 262, the drama flourished better in Anglian than in Saxon England, that is, better in the districts where Homilies and Bible stories were cultivated than in districts where the Legends flourished. On the other hand, the similarity in subject-matter of the Bible stories and the Mysteries, makes it very evident, as is seen from the foregoing sketch, where the elements of early drama are to be sought for.

II. The devil in the York, Townley, Coventry and Chester Mysteries.

General character. — The character of the devil in the English Mysteries is almost entirely serious. This peculiarity is due, not only to the nature of the devil-scenes, which are, in themselves, tragical, but also preeminently to the fact that the devil of the English stage is the creation, not of the people, but of theology. This is evident from the names of the devils, as well as from their speeches and acting; and, as long as this serious conception of the nature of the devil was retained, it was not possible to treat him satirically or humorously. The writers of the Mysteries have, on the whole, remained true to the sources from which they drew their materials, chiefly the Bible and the apocryphal writings. Accordingly, the devil-scenes in the English Mysteries have not made any special developments. It is only in the treatment of some of the under-devils that the authors have freed themselves to any extent from tradition. Examples of this freer treatment of the figure of the devil occur exclusively in the interpolations and revisions, particularly of the Townley and Coventry Mysteries. Here some of the devils become comical and satirical.

The sources of the figures of the devils in the Mysteries. — The greater of the devil-plays and, at the same time, those which are more than all others common to all the cycles, are the *Fall of Lucifer,* the *Temptation of Eve,* the *Temptation of Christ,* the *Harrowing of Hell* and *Doomsday:* all of which are taken directly from the Bible, or rest upon old interpretations of biblical passages, the latter being particularly the case in the *Harrowing of Hell.* For the figures in the remaining devil-plays, the Mysteries go beyond the limits of biblical tradition. A comparison of these plays in the various cycles is particularly instructive, some having devils and some not; for example, in the *Slaughter of the Innocents,* devils appear in the Coventry and Chester cycles, but not in York and Townley; and in the *Ascension of the Virgin Mary* in Coventry, but not in York: cf. Table I.

The sources of the figures for which there is no corresponding biblical prototype are various: 1. The figures of the devil in the *Last Supper* (Coventry), the *Conspiracy of the Jews* (Coventry) and the *Dream of Pilate's Wife* (York and Coventry), owe their origin to the mediaeval idea which regarded the devil as the instigator of the evil actions of men. This idea, so prevalent in the old theological literature, may itself rest ultimately upon such passages as *Luke* 22, 3, where Satan is mentioned as entering into Judas Ascariot, though he does not appear in person, and I *John* 3, 8. 2. The devil in the *Ascension of the Virgin Mary* (Coventry) owes his origin to the underlying thought in the Legends — the irreconcilable feud between the fiend and the saint. In the case of the Virgin Mary this feud is especially emphasized by the peculiar position, which she holds in the Christian system: cf. *Genesis* 3, 15. 3. The figures of the devils in the *Slaughter of the Innocents* (Coventry and Chester), *Antichrist* (Chester) and the *Death of the Virgin Mary* (York), rest upon the mediaeval notions about death as they are embodied in the description of the death of the Virgin Mary in *Cursor Mundi* 20219, seq, and in the famous block-book entitled *Ars Moriendi.*[1]) According to the *Ars Moriendi* devils are accustomed to come to the bedside of the dying, in order to make a last effort to secure the departing soul. The souls of the wicked, as Judas, Herod and Antichrist, fall as a matter of course to the devil: cf. *Cursor Mundi,* 16528.

Origin. — According to the Mysteries, the devils are fallen angels, in whom, as a result of the fall, a great transformation has taken place. This transformation is indicated, first, by the changed names of the devils: in the *Creation and Fall of*

---

[1]) *Ars Moriendi,* editio princeps, 1450? Translated and published by Caxton, 1490 and 1491.

Lucifer (York) „Primus angelus deficiens Lucifer" and „secundus angelus malus" became „Lucifer diabolus in Inferno" and „secundus diabolus". Also in the *Creation* (Townley) „Primus malus angelus" and „secundus malus angelus" become „primus demon" and „secundus demon". Similarly in *Cursor Mundi* 4789, „Lucifer" changes to „Satan". Second, the devils in their wailings refer to their origin and to their present changed condition: cf. Townley, 5/134—138. Again, Lucifer, in a monologue, gives exact information concerning the number of the fallen angels: „The X part fell down with me", Townley, 8/254—7. The cause of the fall, according to the Mysteries, was the arrogance and pride of Satan. This is expressed dramatically in the form of a contention in heaven between the angeli boni and the angeli mali, by the daring resolve of Satan: „I will go sittyn in Goddes se", Coventry, 20, and by the presumptuous demand of Satan that the angels bow before him, Chester, I, 15.

Number: Names. — The number of the devils which appear in the Mysteries, their names and their distribution in the different cycles are given in the following Table I: (v. p. 19).

It is to be observed that the number of the devils is always as limited as possible; many plays have only one, the usual number is two or three, in one play, the *Harrowing of Hell* (York), there are five. The number sometimes exceeded this: cf. Lucy T. Smith, The York Plays, Introduction, XXVIII. The names, Lucifer and Satan, are common to all the cycles; Belzebub occurs in York, Townley and Coventry; Belial in York and Coventry, Titivillus only in Townley; Lightborn[1] (English for Lucifer) only in Chester, Rewfyn and Leyon[2] only in Coventry. Besides these, very common use is made of the simple terms „diabolus" and „demon", often, where

---

[1]) Lightborn occurs *Genesis and Exodus*, E. E. T., p. 7, and in the Legends, Horstmann, *Altenglische Legende,* 1878, p. 139, Leohtberend, in *Aelfric,* Aelf. Soc. I, 10.

[2]) Collier, *History of English Dramatic Poetry,* II, 259, regards Rewfyn and Leyon as allegorical persons, i. e., the personifications of the hatred of the Jews against Christ, that is, of the villainy and iniquity of the Jews. They are, however, also devils: cf. Chester, I, 17 and 84. Weinhold and Schröder both give them in their lists of the names of devils in the old German drama.

Table I.

Play	York	Townley	Coventry	Chester
Creation	Lucifer (diabolus in inferno) (Secund. diab.)	Lucifer (Primus demon) Secund. demon	Lucifer Angeli mali	Lucifer (Primus demon) Lightborne (Secund. demon)
Fall	Satan		Serpens (diab.)	Demon
Innoc.	*	*	Diabolus	Demon
Tempt. of Christ	Diabolus		Satan Belial Belzebub	Diabolus
Last Sup.			Demon, Rewfyn, Lyon	
Council of Jews			Demon, Rewfyn, Lyon	
Dream of Pilate's Wife	Diabolus		Satan Demon	
Harrowing of Hell.	Belzebub Satan Belial Primus diab. Secund. diab.	Belzebub Satan Rybald	Belial	Satan Secund. demon Tertius demon
Death of Mary	unus diabolus			
Antichrist				Primus demon Secund. demon
Assumpt. of Mary	*		Primus demon Secund. demon	
Doomsday	i diabolus ii diabolus iii diabolus	Primus demon Secund. demon Titivillus	Primus diab. Secund. diab. Tertius diab.	Primus demon Secund. demon

[*) Contains no devil].

necessary for the sake of distinction, with the addition of „primus" and "secundus". A number of names of devils are mentioned in the Mysteries, who, however, do not appear on

2*

the stage. Raynal (Chester, I, 84), Balachar and Ragnal (Chester II,174), Anaball, Astarot, Berith and Bel (*Harrowing of Hell*, York and Townley).[1])

The devil names in the Mysteries are thus almost exclusively biblical; names having reference to the outward appearance, to evil and villainous propensities, or names formed to express humor or satire, do not occur. The Mysteries do not contain a variety of grotesque devils' names; nor are the various devils of the Mysteries possessed of marked individuality, as the very frequent use of the simple designation „diabolus“ and „demon“ indicates. The only exceptions to this statement are the names Rybald, Rewfyn and Leyon, and Titivillus, all of whom, however, occupy comparatively but very little space. The tendency thus to enlarge the scope of devil characters in a popular way in the religious drama in England progressed no further.

Hierarchy of the devils. — Definite distinctions of rank and office cannot from the use of the names of the devils be inferred. In a rough way only two classes of devils can be distinguished, an upper and a lower. Lucifer is to be regarded as the chief. The confusion in the use of the devils' names may be clearly shown by the following example:

„Demon: I am your lord Lucifer, that out of hell came,
          Prince of this world and great duke of hell,
          Wherefore, my name is called Sir Satan“,

Coventry, p. 239; here the designation demon, in the rubric, and the names, Lucifer and Satan, in the text, refer of course to one and the same person. Belzebub appears also to enjoy the rank of a ruler. He says of himself: „Whilst I am Prince and principal“ etc. York, 378/111, speaks, however, immediately thereafter of „Satan, our Sire“, 379/117.

The use of the devils' names in any given play in the various cycles is not consistent; for example, Ribald in the

---

[1]) For further references on devils' names see: Weinhold, *Ueber das Komische im altdeutschen Schauspiele, Jahrbuch für Litteratur und Geschichte*, 1865; Karl Schröder, *Redentinerspiel*, 1893; Wieck, *Die Teufel auf der mittelalterlichen Mysterienbühne Fankreichs*, 1887; Osborn, *Die Teufellitteratur des 16. Jahrhunderts, Acta Germanica*, III, 3, 1893.

Townley Plays corresponds, for the most part, to diabolus in York. Belzebub in Townley corresponds sometimes to Belzebub and sometimes to the first, sometimes to the second diabolus, and sometimes to Satan in York. However, when a name of a devil occurs in the text of different cycles, the names agree: for example, Belzebub, York, 377/97; Townley, 296/93, and Rybald, York 378/99; Townley, 296/94.

A system of titles of rank and particular modes of salutation among the devils is only to a limited extent developed. Such expressions as are to be found in the Mysteries are classified in the following

Table II.

a) *Of Satan and Lucifer:*

Sir, York, 382/169; Townley, 298/172; Chester II, 75,81; Coventry, p. 399.

Sir Satan our sire, Townley, 296/111.

Sir Satan our sovereign, sir, Coventry, p. 205.

Sir Satan in the herne, Coventry, p. 399.

Sir Lucifer, lufly of lyre, Townley, 296/113; York, 379/319.

Satan our sire, York, 379/117.

Lucifer, that lord, Chester, II, 175.

Lufly Lucifer .... king and lord of sin and pride, Coventry, p. 207.

b) *In Assemblies:*

Ye dear worthy devele of hell, Coventry, p. 205.

hell hounds, Chester, II, 174.

Hell, hell, Coventry, p. 309.

fellows, Townley, 9/260.

c) *Said of himself:*

A devil most doughty, Coventry, p. 308.

a devil full dark, Coventry, p. 21.

prince of this world-great duke of hell, Coventry, 239.

d) *Said by God or Jesus:*

fiend, York, 478/154; Townley, 301/250; etc.

thou wicked fiend, York, 386/334.

thou foul Satan, Coventry, p. 211.

this foul fiend, York, 183/183.
devil, Townley, 304/357.
thou wared wight, York, 180/73.
warlow, York, 181/115.
wicked worm, York, 27/150.
traitor, Townley, 303/321.
ye princes of hell, Townley, 299/193.
ye princes of pains, York, 379/122.

e) *In ordinary conversation:*
master, Chester, II, 199.
fellows, York, 4/94; 505/217; etc.
master mine, Townley, 372/171.
hell hounds, Coventry, p. 399.
my friend and frery, Chester, I, 17.

f) *Said by the „Tapster“:*
Sergeant, Chester, II, 81.

(g) *The devil designates Christ as:*
that lurdan, York, 179/32.
swain, York, 179/19.
this gentleman Jesus, York, 277/161.
this traitor, York, 381/150.
dastard, York, 382/180; Townley, 299/183.
harlot, York, 383/185.
gadling, York, 384/212.
fellow, York, 388/284.
page, York, 379/125.
Sir, York, 183/151.
belamy, York, 385/213.
thou witty man, York, 180/55.
brodell, Townley, 297/122.
lad, Townley, 297/138.
dossiberde, Chester, I, 201.
stubborn fellow, Chester, II, 77.
popilarde, Chester, II, 76.
this shrew, Chester, II, 75.

h) *God is addressed as:*
god, Chester, II, 196.
goodman, Chester, II, 196.)

Outfit. — Neither the stage-directions nor the words of the players furnish very exact information concerning the appearance and outfit of the devils. It is merely emphasized that they are black, ragged and frightful. The devils bewail, among other things, the loss of their former beauty and brilliancy;

"We that were angels so fair,
And sat above the air,
Now are we waxen black as any coal,
And ugly tattered as a fool",   Townley, 5/134—7.

The stage-direction in Coventry, p. 307, indicates the same, "Here entereth Satan into the place in the most orryble wyse". But just what this "most orryble wyse" was, or how it was attained, is not explained. The expressions "loathest", York, 5/100; "figure foul", York, 478/155 etc. are also used in this same general way.

The account-books of the Guilds are a much richer source of information in regard to the devil's outfit. Sharp in his *Dissertation on the Coventry Mysteries* has given a number of interesting extracts from these books, such as: "Heare for the demons cotts and hose", p. 59, "Devel's face", "head", "malle", "clubbe" and "staff", p. 56. The devils were thus furnished with hairy suits and with masks and were armed with clubs. Satan refers once to his club as a "crocket camrocke", Chester, I, 186, a sort of crooked club made of buckram. Sharp gives also some cuts of the wall-decorations of the Chapel at Stratford-on-Avon, illustrating hell-mouth and groups of devil-figures: See plates 6. and 9. According to Townley, 350/103, the secundus demon very probably did not wear a mask from the fact, as he asserts, that he made faces: "gryned and gnast".

According to the *Proemium* to the Chester Plays, dated 1600, the butchers of the town were charged with the exhibition of the devil in his feathers "all ragger and rent", p. 5. Robert Rogers, Archdeacon of Chester, who wrote in 1609, condemning the "midsomer showe", corroborates the description given in the above mentioned *Proemium*; among other things, he mentions: "ye divil in his fethers before ye butchers, . . . which were reformed and amended": See Furnivall, The Digby Mysteries

XXIII. The amusing account „*of John Andryons in Devyll Apparell*", 1553?, reprinted in Collier's *History of English Dramatic Poetry* II, 263, shows how realistic and effective the make-up of the devil at that time was.

Occasionally the devil disguises himself, but he does this only in the play of the *Temptation of Eve,* where the biblical account requires it. He says in York, 23/231: „In a worme likeness I will wend", and in Chester, I, 27, „the edders cotte· I will put on"; this latter remark is immediately followed by the stage-direction: „Superius volucris penna, serpens pede, forma puella."

The stage. — The Mystery-stage consisted of two, or sometimes three, stages or platforms, one above the other, the lower one of which represented hell: Strutt, *Manners and Customs* III, 130. The stage-direction in the *Mary Magdalene,* The Digby Plays, p. 67: „Here shall enter þe prince of dylles in a stage and hell onderneth þat stage", refers explicitly to this arrangement of the stages; also in Coventry, p. 309: „Here xal a devil spekyn in hell." The entrance to, and at the same time a part of, hell, was the hell-mouth; before which most of the hell-scenes took place. Sharp in his Dissertation gives a number of drawings of hell-mouth, together with some extracts from the account-books of the Guilds concerning the construction of and care for the same, p. 16.

The Fall of the angels was in all probability, represented realistically in the York and Chester Plays. The speech of Lucifer, York, 4/92, refers to an actual fall: „all goes down"; this passage is lacking in the corresponding play in Townley. In Chester, I, 16, God commands Lucifer to fall and the stage-direction provides as follows", Now Lucifer and L[ightborn fall]". In Coventry, p. 21, on the other hand, there is neither a fall nor use of fire: the representation of this scene, in this cycle is decidedly flat; God commands Lucifer to fall, Lu-cifer answers: „Thy will I work, to hell I take my way."

None of the great Mystery-cycles contain, in the stage-directions, any mention of the use of fire. Sharp found in the account-books only one entry for fire in hell-mouth, and that of a late date: „1557—I tm̃ payd for keeping fyer at hell mothe iiij d." The use of fire on the stage was rather the peculiarity of the Digby Plays. In the York Plays, however, Lucifer complains, at the time of his fall, of intolerable heat, „slyke hat“, 5/97, and again, he complains of the heat and of the smoke, which rolls up from below, „ye smore me in smoke“, 5/117.

The scenes in the *Harrowing of Hell* are treated with more than usual detail and are distinguished, especially in the York and Townley Mysteries, by much noise and commotion. The play consists of three parts: First, the excitement among the souls that have been long imprisoned, the astonishment and hasty council of the devils, York, 377/97, Townley, 296/95, the barring of the doors. Second: The bursting open of the doors at the sudden coming of Christ, unusual consternation of the devils:

> „Harro! our gates begin to crak,
> In sunder I trow they go,
> And hell I trow will all to-shak,
> Alas what I am wo!“   Townley, 300/212.

Third: Satan is overpowered and bound by the Archangel Michael, York, 392/340, Coventry, p. 345.

The Occupations of the devils: Councils. — Real councils of the devils are held only twice: just before Satan attempts the temptation of Christ, Coventry, p. 205, and in the *Harrowing of Hell*, Chester, II, 74. In the last-mentioned play, in the York and Townley Mysteries, an assembly of the chief devils is summoned, but, owing probably to the sudden appearance of the Savior, it was prevented. Occasionally a chief devil appears as if addressing a supposititious audience, and thereby announces his plans, Townley, 9/260; Coventry, p. 239. Occasionally the devils take counsel with one another wherever they happen to meet, and discuss their undertakings, for example, in *Doomsday*, York and Townley, and in the *Death of Antichrist*, Chester, II, 175.

Conveying of Souls to Hell. — A special occupation of the devils, as the passage in *Cursor Mundi*, 16528, teaches is the conveying of the souls of bad men, especially of hardened sinners and adversaries of God, to hell. The stage-direction in the *Slaughter of the Innocents*, Coventry, p. 186, is as follows: „Hic dum buccinant mors interficiat Herodem et duos milites subito et diabolus recipiat eos:" the devil enters immediately with the shout, „Alle oure! alle oure! this cattle is mine, I shall bring them into my cell". In a similar way two devils carry off the soul of Antichrist, relate jokingly that they deeply regret his departure, and describe how he shall hang on a hook in hell; the secundus demon says to his mate: „Thou take him by the top and I by the tail", Chester, II, 176.

In the *Death of the Virgin Mary*, York, p. 478, Mary prays to her Son, just as in the *Cursor Mundi* and in almost the same words:

> „And dear Sone whane I shall dye,
> I pray þe þan for þi mercy,
> þe fende þou latte þou nought see", 127—134.

She then prays for sailors, the oppressed, and for women in childbirth, all which prayers Christ grants, excepting the first: „But mother, þe fende must be nedis at þyne ending, In figure foule for to fere þe" (freighten thee), 155, but He promises to send His angels to be about her. At the conclusion of the piece are these words: „Cum uno diabolo", but the diabolus says nothing in the play. The resemblance of this scene to the eleventh picture in the *Ars Moriendi* is striking, there the soul of the dying person is represented rising in the form of an infant into the outstreched hands of the angels, while six horrible fiends rave in disappointment about the bed.

With the living the devils have in the Mysteries but little to do, excepting with Eve and with Christ. Only Rewfyn and Leyon associate exclusively with men; they instigate the Jews against Christ, Coventry, p. 250, 260, and close the bargain with Judas. Later, Coventry, p. 275, a devil, in all likelihood Rewfyn or Leyon, comes to praise and encourage Judas. In the *Dream of Pilate's Wife*, Coventry and York, a devil enters in order to inspire the dream. In the *Harrowing of Hell*, Chester, II, 83, one of the devils assumes a function which, in the English Mysteries, is very unusual; he wants to marry the tapster who is left behind at the time of the delivery of the souls: „Welcome my dear lady I shall thee wed". This episode is manifestly an interpolation: according to Halliwell it is wanting in the Harleian Ms.

In the *Doomsday* the devils that appear are all under-devils, none of the chief devils appearing; in this play they have the task of laying claim to and selecting out the condemned souls. In York, 505/217, and Townley, 370/112, the devils prepare to fight for their so-called „fee“, but the fight is not carried out, a court-or judgment-scene takes its place. The writers of the Mysteries seem hardly able to handle a court-scene successfully, although the English judicial usages were certainly not unknown to them: cf. Miss L. T. Smith, York Plays, *Introduction*, LVII. As a matter of fact, the judge, in this case, God or Christ, is omniscient, and this may account for the weakness of the dramatic interest. In Coventry, indeed, the condemned are already branded as such, the devils are really superfluous. In York, 505/217, seq., this scene occupies only twelve lines, the devils rely simply upon the justice of God in the matter: „He shall do right to foe and friend“, and express themselves as confident that the truth now will out, but at the final condemnation of the souls no devils appear.

In Coventry, p. 403, seq., the lost souls beg for mercy: „No. mercy“, says the devil, and points to the mark on their brows. God utters his judgment and the devils go through the crowd picking out the condemned according to the categories of the deadly sins. They briefly describe each of these sins, but without humor or satire; only lechery is designated by the name of Sybile Schlutte.

In Chester, II, 194, seq., the devils, as in York, rely on the justice of God, with this difference, however, they expressly remind God of his own laws, and refer to the New Testament:

„These words, God, thou said express,
As Matthew thereof bears witness . . . .
And lest thou forget, good man,
I shall mind thee upon,
For speak Latin well I can . . . .
Filius hominis etc. . . .
Therefore, if righteous thou be,

These men are mine,
Or else thou art as false as we" etc.

They then lay claim to various classes of sinners, a pope, an emperor, a king, and a queen. Jesus utters his judgment, the devils declare themselves well satisfied: „A, Sir Judge, this goeth aright", and, with cruel words, turn to the condemned.

The secundus demon in this scene in the Chester plays, whose rôle is apparently an interpolation, appears to regard himself in duty bound to see that nothing is forgotten. In great anxiety he calls out: „Forget not these thieves two!", referring especially to a merchant and a lawyer. In the merchant's case he had kept an accurate account; for every misdeed, the devil had put a kernel of corn into a sack, until the sack had finally become so heavy as almost to brake his neck, Coventry, p. 199—200.

Fear. — As a rule the devils are brave in spite of the fearful situations in which they frequently find themselves. Satan, for instance, orders his armor brought: „Myself shall to that gadling go", *Harrowing of Hell,* York, 384/212, and at the same time he blames the cowardice of the others. Expressions of fear are however not uncommon: „I have great dread", York, 380/137, „Now wax I wode, and out of my wite", York, 292/344, „My wit waxes thin", 296/91, „I gin quake", Coventry, p. 30, „sore afraid, Chester, II, 79.

Cries. — The devils utter, as they enter, preferably the word „harrow", combined frequently with „out", „alas" or „we" etc., „Out harrow" is, in fact, a cry for help, the old Norman hue and cry, and expresses in the plays consternation or pain. It is especially frequent in the plays of great excitement as the *Fall of Lucifer* or the *Harrowing of Hell*; here the use is in keeping with the sense and is justifiable. On other occasions, the devil enters quietly, for example, York, 21/3, Coventry, p. 25, 205, 399, etc., Chester, I, 201 etc. In Coventry, p. 399, occurs the exclamation „Harrow, harrow, we come to town", which is here quite meaningless and indicates a deterioration of the expression to a mere interjection. The devil in this case is happy; he is just leaving hell in order to carry off the soul of the dead prince. This passage is, according to Halliwell, not original: cf. The Coventry Mysteries, p. 418.

The cry „harrow“, or „out harrow“, it is further to be remarked, is not exclusively a cry of devils; it is used by Cain, when attempting to make his sacrifice burn, Townley, 17/275; by Noah's Wife, while being forced into the Ark, York, 48/98; by the King of Egypt in the Red Sea, York, 91/403; etc. etc. Moreover, the devil enters often with altogether different ejaculations: „Make room beleve and let me gone“, York, 178/1; „Ware, ware“, Chester, I, 186; „Alas“, Chester, I, 16. „Anon, master, anon“, Chester, ĺI, 174, etc. etc.

Actual roaring on the part of the devil in the Mysteries simply to make a noise is unusual. An unmistakeable reference to this so-called habit of devils is in the *Harrowing of Hell;* York, 378/99, and Townley, 297/143; Rybald announces the disturbance in the vestibule of hell, Belzebub asks snappishly, „Why roaris thou so Rybald? Thou rorys“. Again in Coventry, p. 399, a devil warns his companions to be quiet, lest „Sere Satan may hear our sound“.

Quarrelling. — As soon as the devils, after their fall, find themselves in hell, they begin to quarrel with each other. The followers of Satan ascribe to him in no gentle words the responsibility for their misfortune. Secundus demon says, „All this sorrow thou hast us sought“, Chester, I, 16. Lucifer, however, rejects the charge with scorn. In York, 5/115, seq., Lucifer and secundus demon, give each other the lie, and not being content with words, they soon come to blows; the devils soundly trounce Lucifer, their leader: „We lurdans have at you“. The remark of Belial, as Michael throws Satan into hell, is anything but sympathetic:

> „þis saide we are,
> Now shale þou fele þy fitte“,   York, 392/345—6.

A subject of dispute between the devils (hell) and Satan was also the question of admitting the Savior, after the crucifixion, into hell. This contention, already found in the description of the *Harrowing of Hell in Cursor Mundi,* occurs as a sort of prelude to the *Dream of Pilate's Wife* in Coventry p. 309; the devil blames Satan for accomplishing the crucifixion of Christ; Christ could, should He come to hell, do much injury to the place; accordingly, Satan hurries to Pilate's Wife and tries thus to prevent the execution.

Scolding. — Satan is occasionally given to scolding. In the *Harrowing of Hell,* (York and Townley) he blames the devils for their weak defence of hell. As Rybald brings the startling news „Limbo is lorn, alas!", Satan curses: „hanged be thou on a crook",

> „Thieves, I bade ye should be bowne,

To ding that dastard down", Townley, 300/217—219. Belzebub objects to these words; he remarks, that is very easy to say, go yourself, and defy the Savior. And because Belzebub is unable to resist Christ's bitter strokes, Satan again severely scolds, Townley, 300/225. In Townley there are a number of verses interpolated, which are lacking in York, namely, where Belzebub in his anxiety calls up Satan; Satan had apparently not been present and appears at first unable to grasp the situation. He answers angrily and brutally:

> „The devil you all to-har
> What ails you so to shout
> And me, if Icome nar,
> Thy brain bot I bryst out", Townley, 797/142—45.

Oaths and opprobrious epithets. — It is remarkable that the devils in the Mysteries are but seldom addressed by each other or by their opponents with oaths or curses; these for the most part fall to the lot of the Savior, owing probably to the fact, that the opponents of the Savior are mostly persons of the lower order, such as servants or lackeys, or are objects of popular hatred such as Herod, whose leading characteristic is profanity; while, on the other hand, the opponents of the devil are exalted personages, God, Christ and Eve, who do not swear at all. A list of the oaths opprobrious epithets used by the devils is given in the following

Table III.

a) *By the devil:*

dewes, York, 4/92.

develeway, York, 380/133.

what devil, etc. York, 385/223, Townley, 297/116.

a develys name, Coventry, p. 390.

The devil yo all to-har, Townley, 297/142.
The devil may speed thy stinking face, Chester, I, 16.
by Belzebub, Chester, I, 26.

b) *By Mahommed:*
Coventry, 199, Chester, II, 197, 199, 200.

c) *Asservations:*
to swear on a book, Townley, 370/100.
by my sovereignty and principality, Chester, I, 201.
mayfay, Townley, 373/188.
by my lewtie, Chester, I, 186.
as I broke my pane, Chester, I, 27; II, 197.

d) *Imprecations:*
high mot he hang, York, 181/117.
hanged be thou on a crook, Townley, 300/216.
The devil you all to-har, Townley, 297/142.
high might you hang right with a rope, York, 178/4.
The devil may speed thy stinking face, Chester, I, 16.

e) *Opprobrious epithets:*
lurdan, York 5/108, Townley, 72/239.
faytour, York, (feature) 80/213, Townley, 298/160.
false faytour, Chester, I, 16.
Thieves, Townley, 300/217.[1])

Obscenity. — It is still more remarkable that the devil of the English stage is almost entirely free from obscenity; with the exception of the really harmless word in Coventry, p. 21 and 30, there is, in the speeches of the devils, hardly any traces of vulgarity or obscenity, even from the modern point of view. To be sure Titivillus and the devil in Coventry, p. 101, 104, border closely, in their satirical remarks, on the obscene. Titivillus uses once the word „luddokys", Townley, 377/314, and speaks of „Nell's . . . smock" which is open behind: Townley 377/328. But, as has already pointed out, these passages are of a later date.

Malevolent speeches. — The devils are accustomed to hold up before their victims the terrors of hell. This trait is likewise

---

[1]) Wanting in the corresponding passage in the York Plays.

characteristic of the later revisions of the Mystery texts. In Coventry, p. 186, for example, the devil tells Herod that he will teach him „Plays fine" and show him „such mirth as in hell". The tertius demon in Chester, II, 83, welcomes the dishonest tapster to „endless bale". The devil in Coventry, p. 275, says to Judas, that he shall have the honor of sitting beside the devil in hell „in fire and stink". In Chester, II, 199, the devil addresses the condemned souls thus, „Judged you shall be to my belly . . . . and delivered be you never". Titivillus and his companions, as they drive the condemned souls off to hell, describe with malicious joy, the pains of hell, the pitch and the messes of rotten oysters. Likewise, concerning the hell-fires and the kindling of the same, the devils are apparently much concerned: „Blow flames of fire to make them burn", Coventry, p. 399; similarly, p. 430, Chester, II, 176, 199.

Malice and hatred. — The devil gives expression occasionally to his hatred of man and to his desire for revenge. In Coventry, p. 29, God asks the devil, why he has led man astray; the fiend replies that he had long cherished great enmity against man because man had been put into the place in heaven which Lucifer and his followers had forfeited: „for I am full of great envy, of wrath and wicked hate;" similarly, Chester, I, 17.

Monologues. — As the devil undertakes the temptation of our first parents, he comes upon the stage in a state of great tribulation, „for woe, my wit is in a war", York, 21/1, describes his bitter lot, and announces his purpose of defeating the plans of God, his motive being to prevent man from enjoying the place in heaven which he himself has lost, York, 22/1, seq., Chester, I, 25, seq. The process is the same in the *Temptation of Christ,* York, 178/1, seq., Chester, I, 201, and in the *Dream of Pilate's Wife,* York, 277/159, seq., Coventry, p. 308. In the *Temptation of Eve* in Coventry this announcement of his plans on the part of the devil is wanting.

It is a marked characteristic of these monologues that they always precede the action to which they relate; that is, except to make some remarks concerning his own fall, or concerning his deeds in the world — matters having nothing to do with the play — the devil explains in advance only his undertakings, but never reports or comments on his own past actions in the play nor on those of others.

The demon in the *Conspiracy of the Jews*, Coventry, p. 239—243, opens the play with a long speech, boasts of his activity in the world, „the dyvercyte of my varyauns", lashes satirically the vices of the times, and describes euphemistically the seven deadly sins. As he makes his exit, he declares himself at all times ready for action, but he does not appear again in the play. This speech has but very little connection with the play in question, for the play deals with the betrayal of Christ and has nothing to do with social conditions in general. See below, p. 34 for a discussion of the relation of this passage with *Judicium* of the Townley Plays.

The devil addresses the audience directly but very seldom. In Chester, II, 176, the devil says to the dead Antichrist, „a doleful look that thou now deal to all this fair company". In the *Slaughter of the Innocents*, Chester, I, 186, the devil threatens the audience with his club, and warns those who are in the habit of giving scant measure with the same fate as Herod, there are many here who may perhaps like to bear him company. With the remark that he will soon return for others he takes leave of the audience.

Asides. — The only examples of asides in the speeches of the devils are in the *Temptation of Christ*; the devil, each time he is repulsed, comments on his failure and announces his resolve to try again, York, 181/85,97, 182/125, Coventry, p. 203, 205.

Titivillus and his companions. — The devil-scenes in the *Judicium, or Doomsday* in the Townley Plays, verses 89—384 and verses 386—532, are interpolations. This is evident from the versification, which differs from that of the rest of the play; moreover, these scenes are lacking in the York Plays, whose *Doomsday* otherwise substantially agrees with that of the Townley Plays. These scenes require special notice, Titivillus et al. differing as they do, in their actions and speeches, from the usual type of the devil in the Mysteries.

As is well-known, the Townley Plays contain several scenes of broad humour, for example, Noah and his wife, Mac and the other shepherds, Cain and his servant, and some others.

Among these comical figures is to be included also that of Titivillus and his companions, which, excepting a few traces here and there, contains the only devil-humour in the Mysteries. Pollard is of opinion that all these comical scenes in the Townley Plays are the work of some unknown but gifted humorist and poet; cf. Townley Plays, E. E. T. XXX. This assumption is contradicted, however, by the condition of the text of the *Judicium*. There are, for example, in the speeches of the devil in verses 225—366 repetitions from the speeches of the devils in verses 143—185,[1]) repetitions which would hardly have been made, were the whole of the work of one author. Further, in the speeches of Titivillus, verses 237—305, are a number of striking correspondences to passages in the speech of the demon in the Council of the Jews, Coventry, p. 69—109.[2])

Action. — The primus demon enters with the usual cry „out harrow", 370/89, the two demons „primus" and „secundus"

[1]) 1. Of lurdans and lyars, 145, repeated 359.
2. Mychers and thefes, 144, repeated 359.
3. Of flytars, of flyars that no man lefys, 146, flytars and flyars that all men reprefes, 360.
4. A bag full of brefes, 143, Of brefes in my bag, 225.
5. Of femynyn gender, 161, Of femellys a quantite, 253.
6. backbytars, 161, repeated 366.
7. fals quest-datyrs, 185, fals dedys forgars, 365.
8. ire, 152, repeated 332.
9. Thou art the best hyne that euer came beside us, 170. Now the best body art thou that euer cam here in. 272.

[2]) 1. long pekyd shoon, Coventry 69; hemmyd shoyn, Townley, 238,
2. thou poverty be chef let pride be present, Coventry, 75. thou prowde as pennyles, Townley 337.
3. Wolle or flokkys, where it may be sowth, To stuff withal thi dobelet, Coventry, 77—78. Of prankyd gownes and shoulders up set, Mos and flokkys sewyd within. Townley, 288.
4. A beggerys dowtere ... to cownterfete a gentylwoman. With here prevy plesawns to get (money) of som man, Cov.101—104. Ilka las in a lande, like a lady nerehande, So fresh and so plesande makys men to folly, Townley, 256—258.
5. I have browth you newe namys, Coventry, 109, (i. e. of the Seven Sins). Yit of the Sinnes seven som thyng speciall now nately to neven, Townley, 305.

vie with each other in grotesque descriptions of their fright
caused by the noise and the blowing of the horn, 370/89—111.
The division of the parts between these two is admirably
carried out, the first serving chiefly to prompt the speeches
and action of the second, very much in the manner of modern
clown exhibitions. Primus demon, although he would rather
walk three times to Rome than go up to this doom, proposes
nevertheless to „make ready our gear", 370/112, similarly
372/175, to „examine our books", 371/140. He then asks of
the secundus demon:

> „Is aught ire in thy bill
> And thou shall drink", 372/153.

„Hast thou aught written there of the feminine gender?"
372/161. The skillful answering of these questions he rewards
with laughter, 372/152, 373/196, and with praise: „Thou are
the best hind", etc., 372/170. The secundus demon, on the
other hand, is the important personage of the scene. He has
with him the books and registers, „a bag full of brefes",
371/143, and „these rolles", 373/183, and is anxiously concerned
to push matters: „Let us go up to this doom up Watling Street",
371/126. He is, however, before all others, the satirist.

Satire. — The secundus demon has in his sack all
sorts of sins, which he pungently describes, enumerating two
separate lists of the vices of the times, 371/142—157 and
373/183—187. He is satirical at the expense of the proud,
371/150, at the expense of the women, the registers of whose
sins fill more books than he can well carry, 372/162—169, at
the expense of the administration of law and justice: „faith
and truth have no feet to stand", 373/188, „the poor people
must pay", 372/189, „worse people worse laws", 373/195, and
finally, refers to the infinite wickedness of the world. He is
of opinion that, had the Day of Judgment been longer post-
poned, it would have been necessary to build an addition to
hell: „We must have bigged hell more, the world is so
warrid", 372/180.

Titivillus. — At this point Titivillus suddenly enters;
„I am one of your order and one of your sons", 373/207: „I was
your chief tollar and sithen court rollar (registrar), Now I am

master tollar", 374/211—213; that is, he has become a Lollard. He relates, further, how he has been instrumental in delivering more than 10,000 souls hourly into hell. Being asked his name, he quotes from a well known poem, „Fragmina verborum Titivillus colligit horum" etc., (Lansd. Mss, 762, see Wright). So far as action goes, Titivillus has but very little to do with the plot. He simply shows his „Roll of Ragmen of the Round Table", 374/224, that is, a list of various delinquents, he describes satirically the new fashions in the dress of men, 374/233—244, and 376/287—295, which he calls „a point of the new get" 376/286, an expression used by Chaucer, *Prologue*, 682, the faults of women, 375/255—271, adultery, 376/277, and a number of political and ecclesiastical abuses, such as the crimes of „false swearers", 376/279.

„raisers of false tax", 376/283,
„kirckchatterers", 376/296,
„gatherers of green wax", 376/284.

The name Titivillus has not yet been definitely explained. Douce derives it from *titivillicum*, a word which occurs in Plautus, Cas. 2. 5; 39, and signifies a trifle, something insignificant; in sense this certainly agrees with the character of Titivillus, whose function is to collect „fragmina verborum". Collier derives the name from totus and vilis. Schroeder, *Redentiner-spiel, Einleitung,* p. 17, regards the word simply as a cloister joke, an anagram of diabolus. The name appears in English, German and French, the oldest occurrence being in the writings of an Englishman, John Bromyard, an opponent of Wyclif. Bromyard was accustomed to collect old stories of monks, some of which he took from Jacob Vitrys, a Frenchman living in the first half of the thirteenth century. Vitrys has an account of a devil who collected the syllables skipped in the services by the monks. Bromyard, in repeating this story, adds the name Titivillus: „Et dixit sanctus: quale nomen habes. Dae-mon respondit: Titivillus vocor, ille autem fecit inde, versum: fragmina psalmorum Titivillus colligit horum" etc.[1]) Bromyard thus appears to assume the authorship of the verses on Titi-villus and the invention of the name.

---

[1]) Bolte, *Der Teufel und die Kirche,* Zeitschrift für vergleichende Literaturgeschichte, 1879: also Wright, *Latin stories,* Percy Society, 1842.

III. The devil in the Digby Plays
(The Conversion of St. Paul, Mary Magdalene)
and Noah's Ark or, the Shipwrights' Play.

The common characteristics of this group are: First, sensation is purposely aimed at. Second: The stage-directions are more explicit. Third: Devil-scenes are introduced or enlarged. The date of the manuscript of the Digby Plays is set by Furnivall at 1480 – 90. The *Noah's Ark* of Newcastle-on-Tyne, is preserved in a print of 1786, it is the only remaining play of a cyclus of 22—23 pieces, known to exist as early as 1426: cf. Holthausen, p. 11, 12.

The sources of the figures of the devil. The third characteristic above mentioned is made evident by the fact that the scenes in question are manifestly interpolations; the entire devil-scene, for example, in the *Paul,* has been inserted by a later hand: See Furnival, The Digby Plays, p. 43. The same is also probably the case in the *Magdalene* : See Ten Brink, *Geschichte der engl. Lit.* II, 320. Furthermore, it is to be observed that the *Legend of Mary Magdalene,* upon which the play is based, makes no mention of devils: cf. the versions in the *Early South English Legends* and in Osbern Bokenhams *Lives of the Saints.* The play of the *Flood* in the York, Townley, Coventry and Chester Mysteries, likewise the biblical account of the *Flood* and *Noah's Ark,* and the biblical account of the Apostle Paul, contain no devils. It is true that, according to the *New Testament,* seven devils were driven out of Mary Magdalene, but this is not sufficient to account for the appearance, in the play, of Satan and other devils. The seven devils reappear in the play as the Seven deadly Sins

Professor Brandl has remarked, *Grundriss,* II, 711, that the construction of the play of *Noah's Ark* approaches that of the Moralities, that is, that humanity here stands between the good angel and the Vice (devil?); similarly again, *Archiv,* 100, p. 436, adding that the rôle of Noah's Wife, as the instrument in the hands of the devil for deceiving Noah, is unusual and reminds one of the methods of the Moralities. Holthausen is of the same opinion; Brotaneck in his edition of this play thinks that the *Noah's Ark* suggests rather the temptation of

Eve by the serpent. This, however, is decidedly too vague;
one could even as well refer to the story of Job, where, indeed, a devil, a woman and a man appear.

Professor Brandl's explanation, that the devil is here introduced in imitation of the Moralities is likewise not conclusive;
if the *Noah's Ark* were written under the influence of the
Moralities, why, one may ask, is the figure in question called
„diabolus" rather than Mischief or Detractio or some
usual Vice-name? Further, the function of this „diabolus"
is not that of the Vice, which is to lead some one into sin
and pleasure; and his manner is not that of the Vice, who,
as tempter, does not use a middle person or agent. For the
devil to act thus through an agent is common, as numerous
examples: the *Legend of St. Madeven*, the *Legend of St.
Dorothe*, the *Blickling Homily on St. Andrew*, E. E. T. p. 239,
etc., etc., prove.

Far more material, however, in this case, is the fact, that
the underlying thought of these plays, including the *Noah's
Ark*, is very different from that of the Moralities. Noah,
Paul, and Mary Magdalene are saints, each being charged
with some great work for God, which the devil, as the enemy
of God, seeks to thwart. Diabolus in the *Noah's Ark* says:

> „Although I have heard say
> A ship that made should be,
> For to save withouten nay
> Noah and his meenye:
> Yet troubled they shall be,
>
> . . . .
> To taynt them yet I trow." 100—106.

Similarly Satan in the *Magdalene* says:

> „In haste we must a conseyll take,
>
> . . . .
> A woman of worship our servant to make." 382—384;

and Belial in the Paul acting upon the suggestion of Mercury,
says with reference to the Apostle Paul,

> ... „this death, doubtless
> It is conspired to reward thy falseness," 486.

Thus it is plain that the persons in these plays and their

grouping is the same as in the Legends; the devils here act from the same motives; the Legend-literature had been and still was much in vogue, and the taste for diablerie was then in the ascendency; on the other hand, the Moralities were yet only in the earlier stages of development. Hence there is every reason for referring the origin of the devils in these three plays to the Legends.

The Miracle Play, *Mary Magdalene,* itself a dramatised Legend, approaches more nearly than any other play of this group to the Moralities because of the prominent part played in the life of Mary Magdalene by the sensual sins. In the Legend, as in the *New Testament,* she is simply represented as a common sinner; the falling into sin, as the result of the temptation of the senses, is the special development of the play and is made the work of the devils:

„A woman of worship our servant to make.“

It is consequently not surprising that this theme, which lends itself so naturally to such treatment, should later be made use of in a Morality, the *Mary Magdalene* of 1567 by Lewis Wager-a play containing a Vice but no devil; the theological literature had, by this time, fallen into decadence and diablerie become out of fashion.

Names. — The names of the chief devils are the same as in the greater Mystery cycles, that is, they are of biblical origin: Lucifer, Satan, Belial, dylle, or diabolus. The term demon, however, does not occur. In the *Mary Magdalene* Satan is designated as „Prince of devils“, p. 68, later as „diabolus“ p. 76, „Rex Diabolus“, p. 82, and „primus diabolus“ p. 82—83. Besides him appears a number of under-devils: „Belfagour“ = „secundus diabolus“, „Belzebub“ = „Tertius diabolus“, and others designated as „tother dylles“.

The Seven Sins according to a stage-direction, *Mary Magdalene,* p. 76, appear dressed as devils: „all the seven deadly sins shall ... be arrayed like VII dylf“, they are expressly called,“ the VII dyllys,“ p. 81, and are under the direction of the Bad Angel, p. 71, 82. Later they are dragged before Satan's court to receive punishment for having let Mary Magdalene escape from their clutches: „Lete þen woman þi bondes break“, 82/732. Their punishment consisted in their being beaten 82/733, drenched with pitch, and, finally, shut into a house and burnt, 83/743. In the midst of all this they are obliged to endure much cursing: Satan calls them „betyll-browed bitches", 82/724, „horesons", 83/744. The devils in the *Paul* are Belial

> „Prince of the parts infernal,
>
> Next to Lucifer in Majesty", 43/412

and Mercury and another devil, Belial's Messenger, called on one occasion, „fool," 44/441.

Outfit : Fire. — The use of fire as a part of the equipment of the devils may be regarded as the specialty of this group. Almost every stage-direction contains some provision for the use of fire or fire and thunder: „Here to enter a devil with thunder and fire, *Paul* p. 43; „Here shall enter Mercury with a firing", *Paul*, p. 44; Similarly, at the exit, „here they shall vanish away, with a fiery flame and a tempest", *Paul*, p. 46; Similarly, *Magdalene*, p. 81.

The oldest stage-direction of this sort in the history of the drama, is in the Morality, the *Castle of Perseverance*, A. D. 1400; it is as follows, „& he that shall play Belial, loke þᵗ he have guñe powd' breññg in pypys ז h's hands & ז h's ers & ז h's ars whañe he gothe to batayl": See Sharp's *Dissertation*, Plate 2. The devil here is thus provided with a sort of fireworks consisting of gunpowder in tubes and carried on different parts of the body. In Heywood's *Play of Love*, A. D. 1533, it may be seen how this was produced: „The Vice cometh in running with a huge tank on his head full of squibs fired", etc., Fairhold, XXVI. Squibs are a kind of fire-cracker. Marlowe has also made use of this method in the equipment of the figure of the devil; in *Doctor Faustus* Mephistopheles introduces a female devil, „with fierworks", text A, 1604. In text B, 1616, this stage-direction has been omitted.

Hell-mouth is an important apparatus of the devil-scenes. The use of a heavier sort of fireworks in the hell-mouth or in hell itself is indicated in *Paul*, p. 43, 46, in *Magdalene*, p. 81. An attempt at the sensational in the use of fire is made in *Magdalene*, p. 83: „Here shall þe tother deylles sette þe house on a fire". Furnivall appears to be in doubt as to what house is meant in this passage, but it is more than likely no particular house whatever; the seven little devils and their leader, the Bad Angel, were simply enclosed in some house, or shed and burnt, indeed, a very appropriate devil's punishment. Satan complacently remarks: „that shall hem a-wake", 83/743, „they shall be blazed both body and hals", 83/744.

Costume. — The plays of this group give but very little exact information concerning costume. In the *Magdalene* the Deadly Sins enter dressed as devils; the stage-direction is, „they shall be arrayed like VII dyll", p. 76, which apparently assumes that the devil's outfit was something well-known, the mention of details, being, consequently, not necessary. In the *Noah's Ark* diabolus swears once by his crooked nose, 123; Holthausen, p. 38, is inclined, consequently, to believe that the devil here wore a mask. This assumption, however, is unnecessary, the probability being that the devil merely uses the expression equivocally, hinting at his real, though disguised, character; for in fact, he appears as a gallant cavalier and is not recognized by Noah's Wife as a devil, cf. Holthausen, p. 19. Noah's Wife uses the word devil more than once, but in each instance merely as an oath and not as applying to the character of diabolus.

Occupation. — The under-devils in this group play only very subordinate parts. Mercury in the *Paul* is the trusty messenger and counsellor of his master; Belfagour and Belzebub in the *Magdalene* assist Satan in punishing the Bad Angel and the Seven Sins. The principal devils of this group associate with human beings in the *Noah's Ark,* with allegorical personages, Mundus, Caro, and the Seven Deadly Sins in the *Magdalene,* or exclusively among themselves in the *Paul.* Their function is, as has already been emphasized, to thwart the work of God's servants and heroes. In the *Magdalene* they plan specially to lead Mary into sin; in the *Paul* they try to accomplish the death of the Apostle, that is, the function in each of these plays is the same as that of the devils in the Legends.

In the *Paul* Belial is found anxiously waiting for the return of his messenger, Mercury, Mercury finally brings the news of Paul's conversion and baptism, there upon the two devils set up a wail, p. 45. Mercury suggests, however,

that this is all useless; the high-priests must be instigated
against Paul, and the death of the Apostle must be accomplished in some shrewd, secret way, but what they do to
accomplish this end is not related.

In the *Magdalene,* the devils, at the suggestion of Satan,
hold a great council to discuss measures for the corruption
of Mary Magdalene: „A woman of worship our servant to
make", 68/384; Satan gives the Seven Sins orders for carrying
out this plan and, in a happy frame of mind, takes his leave,
76/560. He appears again and summons the council of the
devils for the purpose of punishing the Bad Angel and the
Seven Sins, p. 82. The diabolus in the play of *Noah's Ark*
is without any companion. He sets out to thwart God's plan
for saving Noah and his family:

„Yet troubled they shall be...

To taint them yet I trow". 103—106.

To this end he goes to Noah's Wife to plot against the good
old man.

Crying and Howling. — Another specialty of this group
of plays is the roaring on the part of the devils. The outcry,
„Out harrow," is in the old sense only partly retained. In
the *Magdalene,* 87/722, and 91/963, the cry expresses consternation and disappointment, as in the other Mysteries. This
use of the expression is, however, in the Digby Plays, exceptional.

That the outcry is used without special meaning is shown
by the fact that the devils, even in time of greatest tribulation and disappointment, expressly call attention to the fact
of their roaring, as in the *Magdalene,* 91/963: „I may cry and
yell". The stage-direction in the *Paul* provides especially
that the devils howl: „cryeing and rorying", p. 44, „Here they
shall rore and cry", p. 45: cf. „I roar" in the Marco Morality,
Wisdom, 150/225.

In the *Paul,* 43/412, Belial enters with the cry „Ho, ho,
behold me". This is the first occurrence of this outcry; it is
a perfectly natural outcry, but has no specific signification
except to attract attention, or make a noise: cf. further the
shouts of Curiosity: „Hof, hof, hof," *Magdalene,* 73/491; the
devil in the *Disobedient Child,* A. D. 1550?, Dodsley's Old

Plays, II, 304, enters with: „Ho, ho, ho". The statement of Sharp, p. 58, is misleading; he says, „A frequent exclamation used by the demons of our ancient Mysteries was „Ho, ho", and refers to *Gammer Gurton's Needle.* Sharp's mistake consists in applying the customs of the late 16th century to those of the 15th. The customs in question could in the meantime have completely changed. Ben Jonson was also accustomed to send his devils onto the stage with this so called devil's bluster. The character of the devil as a stage-figure had, however, since the middle of the 16th century materially deteriorated; see below, p. 48, seq.

The outcry of diabolus in the *Noah's Ark,* as he makes his entrance, is unique, he says, „Put off, harrow", etc. Holthausen in a note to this passage, verse 93, says that he does not understand it, likewise Brotanek in his edition of the play, *Anglia* XX, 188, is unable to explain it. The explanation, however, is simple enough. The play, *Noah's Ark,* is, as we know, a shipwrights' play; the devil is, accordingly, adapted to his surroundings, that is, he is a sailor. He is found getting into a boat in order to go to Noah's Wife and gives the order to the rowers, „Put off", that is, push off from the moorings. This common use of the term „put off" is, in fact, given in *Muret's Dictionary* under „put"; although Holthausen failed to find it. The further „Harrow" etc., diabolus says complainingly to himself.[1])

Humor. — Traces of humor, mostly humorous comparisons or epithets, are to be found in the speeches of the devils of this group: „I tremly and I trot," *Magdalene,* 76/555; „Flat as a fox", 82/730, is said by the Bad Angel as he appears before Satan, meaning that he prostrates himself in the manner of a fox 'possum; „This hard balys on thy buttocks shall bite", 82/735 and „Scour away the itch", 83/737, are the expressions Satan uses when he orders the eight evil spirits to be beaten. „And that shall hem a-wake", 84/743, is Satan's

---

[1]) Cf. Hickescorner's outcry „Shoot off", 16; he was himself a sailor. Further, the devil appears in a Spanish Morality *„The Journey of the Soul"* by Lope de Vega, as a ship's captain. Ticknor, *Hist. of Spanish Literature,* 1849, II, 160.

remark at the burning of the delinquents. („Bleared is our eye", 92/985, is probably not humorous).

The devil happy. — A happy, satisfied and self-congratulating devil is a new phase of this character. He is proud, „pricked in pride", *Magdalene,* 68/358, he boasts of his royal living, „as a king royal I sit at my plesauns", 63/361. At the news that Mary Magdalene is fallen, he is beside himself with joy, „A, how I tremly and trot for these tidings! . . . for of her all hell shall make rejoicing", 76/555, 9. His rejoicing, however, is soon turned to gall; when he hears of Mary Magdalene's conversion, he is beside himself with rage: „A, out, out, and harrow, I am hampered with hate", 82/722. It is only while engaged in punishing the bad spirits that he, to any degree, recovers his complacency: „Now have I a part of my desire", 83/740. The process is very similar in the case of Belial in the *Paul.* Belial brags of his achievements, but as before, his happiness is of short duration; Paul in spite of him has been converted.

Monologues. — The devils of this group make little or no reference to the audience. Diabolus in the *Noah's Ark* addresses the company directly and expresses the wish that all, who do not believe in him, may be permitted to enjoy the fires of hell in company with Dolphin. With this the piece closes. In the *Magdalene* a dylle (Satan) makes a long complaint, at the close of which he takes leave: „I tell you all in fine to hell I will go", 92/992. This monologue, *Magdalene* 91/963—992, is, so far as the subject-matter is concerned, of great interest, because Satan here, as also in *Wisdom,* represents, to some extent, the ancient chorus; Satan in his speech reports and comments on a number of past occurrences, (the resurrection and the journey of Christ into hell,) which must be thought of as happening between scene 20 and scene 22, but which have not been presented on the stage.

VI. The devil in the Moralities.

Group a. In the early Moralities, *Castle of Perseverance, Mankind* and *Wisdom,* that is, in those which were written before 1500, the devil appears as a constant figure, but, it is worthy of note, he is not the sole representative of evil; there are here associated with him several Vices, including the Deadly Sins. The position of the devil in these plays is, however, one of primary importance, and is, furthermore, excepting in Mankind, always serious. The relation of these several figures, of devil and Vices, to one another can best be shown by means of a short account of the part played by the devil in the plays of this group.

1. In the *Castle of Perseverance*, (about 1400) the devil is Belial. Here he forms, together with Mundus and Caro, that famous trinity which once so largely occupied the minds of theologians. Concerning Belial's rôle in this piece one can say but little, for the reason that we have, as yet, no complete accessible text. According to Collier, *History of Dramatic Poetry*, II, 283, Belial maltreats his accomplices, the Deadly Sins, at whose head he leads the attack upon the *Castle of Perseverance*; he calls them „harlots", curses them „by Belial's bones", and, according to the stage-direction, „et verberabit eos super terram", he beats them because they had let Humanum Genus escape.

What we know concerning Belial's make-up, namely, the fire-works, has already been discussed, see page 40.

2. *Mankind*, (about 1450 — Brandl, *Quellen*). *Titivillus and his rôle.* — Titivillus is summoned with music by the other Vices, 438. He and the chief Vice, Mischief, seem to be on terms of complete understanding with each other, 486, 646, although they do not appear at any time together on the stage. Titivillus, after promising to avenge the three minor Vices, Nought, Newguise, and Nowadays, against Mankind for the beating them with his spade, 487, sends them out to seek for opportunities of doing mischief, 488. Unseen by Mankind, he whispers evil suggestions in his ear while he sleeps, or while he is praying, 545—546, 579—583. He announces to the audience his plans, 511—526, 542, 577—588, and steals Mankind's spade, 535. He reports his previous doings: „Mankind was besy in his prayere etc." 511, „I have brought Mankind to mischief and shame", 592. He praises his own cleverness, 541, 588, and, as he goes out, he takes leave of the audience: „Farewell, everyone, for I have done

my game", 591. In the other principal scenes of the play, the „Court of Mischief" and the scene where Mankind attempts to hang himself, Titivillus is not present.

The Comic Element. — The comic quality of Titivillus comes to light in various ways. As he enters he makes the naïve remark „I come with my legges under me", 439. Another joke is that which he whispers to Mankind, namely, that „þe deullys dede", 579. He attempts to borrow a penny from the Vices, 464, but they have the best of him; they say they have no money, although they had just taken up a collection from the audience on the plea of exhibiting the great Titivillus: „ellys þer xall no man him se", 443, furthermore, he uses scraps of Latin, 460, 462, 564, a few obscene expressions, 546, 554, and blesses the Vices with his left hand, 508.

The outfit of Titivillus consists of a net, 516, and, in all probability, a mask, „hede", 446.

Hostility to the Good. — Titivillus is really not at enmity with the allegorical figure of the Good, here called Mercy: he will only separate Mankind from him: „The good Man Mercy shall no longer be his guide", 513. He furthermore attempts to stain Mercy's character by calling him a horse-thief, 585.

The Temptation of Mankind. — In this play the temptation is effected, as Professor Brandl remarks, by the means of evil suggestion and taking by surprise. This motif, the whispering evil suggestions, occurs also in the *Dream of Pilate's Wife* in the York and Coventry Mysteries, and, is as we have seen, frequently made use of in the Legends, as in the story of *St. Madwen.* It is interesting also in this connection to compare the numerous old drawings in which the devil is represented as whispering in the ear of men and women.

The placing of a board under the surface of the ground in order to aggravate Mankind at his work and to make him „lose his patience", 522, the suggestion of religious pride and the giving him the belly-ache, are motifs peculiar to this play; but that which he suggests to Mankind is the usual means of seduction, namely, that Mercy is an unreliable character: „Trust no more on him, he is a marred man", 586, that tilling the ground, that is, honest labor, is a hard lot, that he should go to the inn, and that he should get him a sweetheart: „and take þe a lemman", 590.

3. The play of *Mary Magdalene,* (about 1480—90, Furnivall The Digby Plays) has already been discussed, see above, p. 37, seq., should be mentioned again in this connection on account of its being a mixture of the Morality and Mystery types. In this piece the evil originates from Satan alone. Satan conducts the conference of the devils and commissions the Bad Angel and the Seven Deadly Sins to debauch the saint. Later in the play he punishes his agents severely, and that for the very same fault for which Belial does his in the *Castle of Perseverance,* namely, for letting Mary Magdalene escape. Before leaving the stage he comments on his own past actions.

4. *Wisdom* (about 1480—90, Furnivall, The Digby Plays) is a Morality constructed according to the plan of the greater Temptation-Mysteries of Eve and of Christ. Lucifer enters in the usual manner of the devil of the Mysteries, represents, however, as regards his activity, the Vice of the Moralities. For example, he announces his plans, 325, 380, he comes for the express purpose of leading man (Mind, Will and Understanding) astray, he speaks directly to him, he slanders the Good, and, in the end, comments, in the manner of the ancient chorus, boastingly of his previous doings: „Of my desire now have a some", 520, „Reason I have made both deaf and dumb, Grace is out and put a roam", 524—525. Just before the conversion of Man he makes his exit.

Outfit. — In order to hide his horrible appearance Lucifer disguises himself according to the stage-direction, thus: „Entereth Lucifer, in a devely array without and within as a proud gallant", p. 150, see also p. 151.

The Comical Element. — The comical element is entirely wanting in this piece except at Lucifer's exit, where we find the following direction, „Here he taketh a shrewd boy with him and goes his way crying", p. 158. This, however, does not at all correspond with Lucifer's rôle in this play and is certainly to be regarded as an interpolation.

Hostility. — Lucifer is, in consequence of the fall, very bitter, revenge is the sole motive of his action:

„Out harrow, I roar,
For envy I lore", 525.

The devil in the *Temptation of Christ* in the York Plays gives
expression to exactly similar motives. The object of the
devil's hostility is, in these cases, especially Mankind: „Man
whom I have in most dispight", etc., 338. Both these traits
— revenge towards God and hatred towards Man — are thoroughly devilish and not vicious, that is, pertain to the devil
and not to the Vice. The preachers, as representatives of the
Good, are however, also the object of Lucifer's hatred, as is
always the case with the Vice: „They flatter and they lie...
there is a wolf in a lambe skin" — 490.

The temptation of Man. — In order to lead humanity
astray Lucifer resorts chiefly to the use of arguments, wherein
he distinguishes himself as a well-schooled logician; his arguments are: 1. There is time enough both for the worldly life
and for the spiritual, as is to be seen, for example, in the
case of Martha, who pleased God, and also in the case of
Christ, who led the so-called „vita mixta", 401, 428; 2. The
religious life is dreary: „they must fast, wait and pray", etc.,
433; 3. The worldly life is not to be despised:

„Behold how riches destroyeth need,
It maketh a man fair him well to feed,
And of lust and liking cometh generation", 458—460.

„What sin is in meat, in ale, in wine?
What sin is in richesse, in clothing fine", 473—475,

and finally he challenges Humanity, in the manner of the
Vice, to lead a jolly life: „Leave your studies ... and lead
a common life", 472, „Leave your nice chastity and take a
wife", 476, „Ever be merry; let revel rought", 505, without,
however, taking part himself therein, as does the Vice; this
is an important distinction between the two figures.

Group b. The appearance of the devil in the remaining
Moralities can be briefly disposed of. *Nature* (about 1500, Brandl,
Quellen) is the first Morality without a devil, and, from this
time on, the devil appears only occasionally on the stage, and
that mostly in the later Moralities. The position of the devil
in the plays of group b is always subordinate.

[1. *The Necromancer,* (printed 1504, described by Warton, *History of Engl. Poetry,* II, 360, see also Skelton's *Works,* ed. Dyce, I, XCIX). This lost piece of Skelton's should be mentioned here on account of its peculiar portrayal of Belzebub, who is the chief character of the play. He kicks the Necromancer, who comes very early in the morning to summon him to court; the plot is the trial of Simony and Avarice, probably Vices; Belzebub plays the judge; the closing scene gives us a view into hell, Belzebub and Necromancer dance for the benefit of the audience, Belzebub trips Necromancer up. The stage-direction: „Enter Belzebub with a beard" is unusual because of its explicitness].

2. In *Lusty Juventus* (about 1550, Dodsley) there appears a devil called simply „the devil". He is dressed as a swine, he bewails the progress of the Reformation: „Oh, oh, all too late" etc., p. 62, and makes known his intention of infecting Youth. This he will do through his son, Hypocrisy, the Vice of the play, whom he calls up and to whom he explains his plans; he promises Hypocrisy his blessing, should Hypocrisy succeed in sowing discord between Youth and Knowledge; he then changes the name of the Vice to Friendship and makes his exit.

3. The *Disobedient Child* (about 1550, Dodsley) contains the devil unaccompanied by a Vice. The devil enters only once and with the cry: „Ho, ho, ho, what a fellow am I", p. 307, he makes a long speech on his power in the world and on the means which he intends to use in the destruction of a certain rich man's son; and these are none other than the Deadly Sins, who, however, do not appear in the play.

4. *Like Will to Like* (printed 1568, Dodsley) represents Lucifer in a decidedly fallen condition, that is, as a target for the jokes of the Vice; he enters dressed as a bear, at whom the Vice at first is greatly frightened, he carries on his breast and on his back placards containing his name „Lucifer" in large letters. He calls the Vice, Nichol Newfangle, his son, and commissions him always to join „Like to Like". The Vice does this with a vengeance, in that he introduces Lucifer to Tom Collier, a clown; the three then sing and dance for the benefit of the crowd. It is especially to be noted that

this is the only instance of the devil singing, while, on the
other hand, singing is almost a constant trait of the Vice; at
the conclusion of the song Lucifer requires the Vice to salute
him as his chief. Nichol, however, perverts the formula of
the salution most fearfully. Lucifer then leaves the stage,
p. 317, enters again, however, shortly before the end of the
piece in order to carry off the Vice on his back. The Vice
mounts him as he would a horse, p. 356.

5. In the *Conflict of Conscience* (1567—70? Dodsly) Satan
opens the play with a long monologue, „High time it is for
me to stir about". He expresses his hatred of Christ, calls
himself a prince of this world, and declares the pope to be
his son and darling to whom he has given the rule in this
world. Further, he explains fully how he tempted Eve, misled
the children of Israel and tempted Christ. In these days he
says the pope has secured two great champions, Avarice and
Tyranny, the minor Vices of this play, who, as it appears,
have not succeeded in defending the pope's power against
the onslaughts of his enemies. Satan will, accordingly, send
him Hypocrisy, the chief Vice of the play: „No sweeter
match can I find out than is Hypocrisy". With an allusion
to his own horrible appearance, he then takes leave of the
audience: „For none will be enamoured of my shape I know,
I will, therefore, mine imps send out from hell their shapes to
show". It is to be noticed here that Satan is not called up
by anyone, and that he comes in contact with no one.

6. In *Appius and Virginia*, (about 1563, Dodsley), there
is evidently a devil behind the scenes; Haphazard, the Vice,
as he enters, is in conversation with him: „Very well sir,
very well sir, it shall be done", and there upon utters the
proverb, „who dips with the devil, he had need have a long
spoon", p. 117—118.

7. In *All for Money* (printed 1578) Satan enters with
the woful words, „Ohe, ohe, ohe, ohe, my friend Sin" (Bii).
The Vice, Sin, however, flouts him, calls him „snotty nose"
and threatens to diminish his kingdom. This is what the
devil dreads most of all and is the reason why he howls so:
„Here Satan should roar and cry"; Satan begs: „My friend
Sin, do not leave me thus", but in vain. Satan then summons

his sons, Gluttony and Pride, and these succeed in winning Sin over, whereupon Satan becomes extraordinarily happy and dances about for joy; Sin, however, will have a reward for his constancy to the devil and demands either the devil's tail to make a flapper of or his mask („face"). The devil cannot spare either but is willing to appoint Sin chief officer of hell; whereupon he makes his exit.

The stage-directions indicate Satan's outfit only in a general way: „Here cometh in Satan the great devil as deformedly dressed as may be"; this reminds one of the ragged appearance of the devil in the Mystery Plays. As already indicated Satan had, moreover, a tail and a mask. In this piece Satan enters only once, and, as it appears, chiefly to give point to the joke, the devil and Sin have fallen out.

The rôle of the devil in the Moralities. — The office of the devils in the Moralities is, on the whole, limited to one thing, namely, that of giving their agents, the Vices, their hellish commissions, for example, in the *Mary Magdalene*, Lucifer sends the Seven Deadly Sins out to debauch the saint. In *Lusty Juventus* the devil calls up Riot, the Vice, and commissions him to counteract the work of the Reformation; Satan, in the *Conflict of Conscience* does the same, and according to the passage in *Appius and Virginia* it would appear that the action of the Vice in the play were, to a considerable degree, dependent, for its start, upon such prompting of the devil.

Among those already mentioned are three figures of the devil which require particular attention, namely, Titivillus in *Mankind,* Belzebub in the *Necromancer* and Diabolus in the Mystery *Noah's Ark*. These, in their principal characteristics, differ decidedly from the rest of the figures of the devil. In the portrayal of these three, the authors appear to have freed themselves from the trammels of biblical tradition. This tendency to treat the devil, as for instance Titivillus, in a free way and to associate him with human beings other than saints or to give him a rôle taken from daily life, as is the case with Belzebub, who plays the rôle of judge, or, as is the case with Diabolus, who is a sailor, indicates a new phase in the development of this figure; but this development has not proceeded further in this direction.

In the Moralities as well as in the Mysteries the characteristic traits of the devil have, on the whole, not been enlarged upon. The English drama shows no differentiation of the character of the devil as an exponent of the comic, of satire, and of carricature, as is the case with the German and the French drama. That is to say, the English does not show a number of devils, excepting the three above mentioned, each with a specific property or commission, designed to take off or stigmatize human faults: as for example, a family devil, a court-devil, a devil of drunkenness, a devil of fashion, etc. etc.: cf. woodcut of the title-page of the *Theatrum Diabolorum,* 1569—75, given in Osborn's: *Die Teufel-Literatur des sechzehnten Jahrhunderts.*

The conception of the devil has remained uniform in England, hence the limited number of devil's names in English. In many plays he is called „the devil“ or has some one of the biblical names Satan, Belzebub, Belial, Lucifer; there are, on the other hand, according to Weinhold about 68 distinct devils' names in the old German drama, p. 18; and according to Wieck about the same number in the French Mysteries p. 7. This remarkable difference between England and the continental countries is to be explained by the fact that the English have used the allegorical figure of the Vice as the representative of the sins and weaknesses of men, while the Germans and the French, who had no Vice-figure, used for this purpose a differentiated devil-figure. Regarding the indigenous character of the Vice-figure see Douce, I, 467, and Ward, I, 160.

The devil negatively considered — It is eminently appropriate, at this point, to make a few negative observations on the character of the devil, in order to settle once for all some questions as to what the devil does not do. Such a process will not only give a more definite conception of the character and function of the devil, but will, at the same time, brand as false a number of very prevalent opinions on the subject: — 1. The devil of the Moralities, on the whole, does not come in contact with the human characters. 2. The comical element is almost entirely wanting, or is at least very weak; he is never represented as stupid or as being imposed upon. Singing, dancing, eating, drinking, in short, making merry, is just what the devil does not do. 3. There exists no definite relation whatever between Vice and devil; it is not the devil's part to offer himself as the target of the Vice's teasing nor finally to drag the Vice off to hell. 4. Except in the earliest Moralities, the devil takes only a very subordinate part in the plot. 5. The devil does not represent human characters.

As exceptions to one or more of these rules are to be mentioned four representations of the devil: 1. Lucifer in the *Mary Magdalene,* is the principal character; he conducts the entire action of the play, excepting, of course, the conversion-scene, and comes in direct contact with human beings. 2. Titivillus in *Mankind* organizes to some extent the action of the play, and is comical. 3. Belzebub in the *Necromancer* is the chief person of the play; he represents a human character, namely, that of judge, he dances, and beats the other characters. 4. Lucifer in *Like Will to Like* is stupid, dances and sings.

Part II. The Vice.

I. The Vice-dramas: Limits of the present investigation.

The Vice-dramas are those dramas which contain a Vice-figure, they are either Moralities or Tragedies, that is, serious plays. The Vice, indeed, is a characteristic feature of the Moralities, the only Moralities not having a Vice being the Moralities of Death and Judgment, such as, *Everyman.* In the pedagogical Moralities, the three so-called *Wit-plays,* the character Idleness may also be considered a Vice, indeed, in the latest of these plays, A. D. 1579, Idleness is expressely called the Vice, but at that late date the designation means but little. Idleness is, in name, a Vice, so also his (or her) accomplices, Ignorance and Tediousness, but the character, even in these plays, is quite subordinate, and, as a Vice, is not well defined; on the whole, Idleness appears rather to be a special development of the *Wit-plays* and quite independent of the Vice proper.

The Vice-Tragedies are three in number; the Vice-figure in these plays is, in all probability, borrowed from the Moralities. The figure of the villain is a type distinct from the Vice in its origin and partly also in its function. Tragedies with figures of the villain-type are, at present, left out of account.

Some of the early Comedies have figures that are in some respects similar to the Vice; Diccon, the Bedlam, in *Gammer Gurton's Needle,* a mischief-maker pure and simple, or Conditions in *Common Conditions* a mischief-maker and buffoon. Such figures however, are quite special and are only partially Vices, they may have been influenced to some extent by the Vice-figure, but can hardly be regarded as factors in the development of the Vice.

The following table gives the dramas chiefly discussed in these pages, together with the characters to be considered. The Vice-dramas may be conveniently classified in two groups, the Moralities and the Tragedies; of the Moralities there are further four main groups, the early Moralities, 1400—1500, the Moralities of the middle period, 1500—1560, the political or controversial Moralities, and the late Moralities, after 1560. This classification is based upon the structure of the Vice-rôle in each period.

Table IV.

	Play	Man	Good	Vice	Minor Vices	Devil
1. Early Moralities (1400—1500)	Castle of Perseverance (1400)	Hum. Genus	Deus	Detractio, Stultitia	Deadly Sins, Bad Angel, Mundus, Caro	Belial
	Mankind (1450)	Mankind	Mercy	Mischief	Nought, New-guise, Now-adays	Titivillus
	Mary M. (1480—90)	Mary M.	Good Angel	Sensuality	Deadly Sins, Mundus, Flesh, Curio-sity	Satan, Devils, Bad Angel
	Wisdom (1480—90)	Mind, Will, Understand-ing	—	—	—	Lucifer
	Nature (1500)	Man	Reason	Sensuality	Deadly Sins, World, W. Aff., Priv.Co. Bod. Lusts	—

	Play	Man	Good	Vice	Minor Vices	Devil
2. Moralities of the middle period (1500—1560)	W. and C. (1506)	Manhood	Conscience	Folly	[World]	—
	Hickescorner (1509)	Freewill, Imagination	Pity	Hicke-scorner	—	—
	Four Elements (1510)	Humanity	Stu. Desire, Experience	Sensual Appet.	(Ignorance)	—
	Lusty Juventus (1550)	Juventus	Good Counsel, Knowledge	Hypocrisy	Fellowship	Devil
	Youth (1554)	Youth	Charity	Riot	Pride	—
	Nice Wanton (1560)	Barnabas, Ismael, Dalilah	Barnabas	Iniquity	—	—
3. Late Moralities (after 1560)	Trial of Treasure (1567)	Lust (Sturdiness)	Just, Sapience	Inclination, the Vice	Greedy Gut Elation	—
	Mary M. (1567)	Mary M.	Christ	Infidelity, the Vice	Pride, Carnal Concup., Cupidity	—
	L. W. L. (1568)	[clowns]	—	Nichol New-fangle, the Vice	—	Lucifer
	Tide (1576)	[Types]	—	Courage, the Vice	Greediness, Help, Profit, Furtherance	—
	Money (1578)	[Types]	—	Sin, the Vice	Pride, Gluttony, Pleasure	Satan

	Play	Man	Good	Vice	Minor Vices	Devil
4. Political or controversial Moralities	King John (1548)	(John)	(England)	Sedition (the Vice)	Dissi- mulation	—
	Res Publica (1553)	(People)	(Respublica)	Avarice, the Vice	Adulation, Oppression, Insolence	—
	Confl. of Conscience (1563)	(Philologus)	—	Hypocrisy	Tyranny, Avarice, Suggestion	Satan
	King Darius (1565)	—	—	Iniquity, the Vice	Importunity, Partiality	—

	Play	Vice
Tragedies	King Cambyses (1561)	Ambidexter, the Vice
	Appius and Virginia (1564)	Haphazard, the Vice
	Horestes (1567)	The Vice [clowns]

II. The Vice-rôle.

Origin. — The question concerning the origin of the devil has been variously answered; it is often maintained that he was simply borrowed from the Mysteries: cf. Collier, II, 262, Ward, I, 60, and others. This theory presupposes a sort of direct, conscious borrowing of a figure or motif, because of some predilection, from one species of literature for another, but overlooks the fact that such borrowing is a mechanical process more peculiar to an eclectic age like the present than to a period preceding the renaissance. That an author of an early Morality ever said or thought, go to, let us make a play with a devil in it after the manner of the great Mystery Plays, is not probable. There were, indeed, at that time, a set of people, who were, perhaps, capable of fulfilling the conditions of the above mentioned theory, namely, the minstrels or jugglers; but, so far as we know, the English drama was never at any time, wholly or in part, in their hands: cf. Ward, I, 16, though perhaps otherwise in Germany: cf. Freytag, 18, 459. Weinhold's explanation that the devil was introduced into the drama because he was the punisher of vice and the father of sin, does not sufficiently cover the ground.

This question is a far more general one and includes the origin not only of the devil but of all the evil characters in the Moralities; the answer, to be of any value, must be based upon the historical facts in the case, so far as these are accessible. The explanation of the Seven Deadly Sins, the Vice and the devil in the Moralities is one and the same, and is not far to seek; indeed, it can be shown not only why the devil and these other figures were introduced, but the sources from which they were drawn, can also be ascertained to a certainty.

The reason for the introduction of the evil figure, or figures, lay in the very purpose of the Morality, namely, the allegorical representation of the conflict between the powers of good and evil over the human soul; the source of these figures was the general traditions of the age. This conflict was, as we know, a frequent theme of discussion in the religious, didactic literature of the Middle Ages; the usual method of treatment was to assume real representatives of each power; these were, on the side of evil, the Deadly Sins and devils. That the Moralities drew their material and inspiration from a common source with this literature, is manifest.

The representatives of the Evil. — As already intimated, the devil was far from being the only representative of Evil; up to the beginning of the 16th century the rôle of the Evil in the Moralities was almost always divided; on the other hand subsequent to the beginning of the 16th century this rôle was concentrated in a single figure, the Vice, or at least in one chief Vice and several minor Vices. Thus in the *Castle of Perseverance*, A. D. 1400, there are one devil, two Vices, and ten minor Vices; in the *Mary Magdalene*[1] (1480—90), the same; *Mankind* (1450) has one devil, one Vice, three minor Vices *Nature* 1500, has one Vice and eleven minor Vices; *Magnificence*, 1515, has seven scarcely distinguishable Vice-figures. The relation of these figures is shown graphically in table IV. The extent of this surprising splitting up of the Vice-rôle, and the relation of the various figures in question, can be easily seen by a brief review of the earlier Moralities.

1. The *Castle of Perseverance*. — Besides the devil, Belial, there are here two undeveloped Vices, Stultitia and Detractio, the latter of whom appears also, at least in name, in the Coventry Mysteries. He figures here as page to Humanum Genus and procures him the acquaintance of Avaritia; the other, Stultitia, appears later in the *World and the Child*, A. D. 1509, as a full-fledged Vice. The minor Vices of this piece are the Seven Deadly Sins, among whom one, Avaritia, plays quite independently a good Vice-rôle; he calls up the other Vices and succeeds finally in leading Humanum Genus, now an old man, astray.

2. *Mary Magdalene*. — The Miracle Play, *Mary Magdalene*, contains an abundance of devils and Vice-figures, but no one chief Vice. After the council of the devils, the Bad Angel, the World and the Flesh, the Seven Sins and the Bad Angel besiege the castle of Magdalene, p. 71. Lechery enters the castle, flatters Mary, tells her not to grieve for her father and finally takes her to a tavern in Jerusalem; a gallant, Curiosity, enters, p. 73, he is represented as longing for a pretty tapster to talk to, and as admiring his attire. Lechery suggests to Mary that „this man is for you“, 507, Mary asks at once to „call him in“, which the taverner does. Curiosity proceeds, without further circumstance, to make love to Mary, asks her to dance and to drink some sops of wine. They then go out together.

The character Curiosity is but very slightly developed, his part in the action of the play very limited; he is neither hellish nor vicious; the part of Lechery, the real tempter in this case, is far more important, she forces her way into the castle and by means of flattery succeeds in leading the victim astray; neither the Gallant nor Lechery is humorous. The expression „höllischer Hanswurst", as used by Professor Brandl, Grundriss, II, 705, does not apply very well to either of these figures.

3. *Mankind.* — The Vice-rôle in this piece is, apart from Titivillus, divided among Mischief and Nought, Newguise and Nowadays; the chief Vice, Mischief, enters twice: he is absent, however, during two of the principal scenes, that of the secondary Vices and Mankind, 68—152, and that of Titivillus and Mankind, 442—624. He returns finally to lead Mankind, who in the meantime has been led astray by Titivillus, further into a life of dissipation, and with this object in view, admits him into his „Court of Mischief", 625, seq.

The accomplices of the Vice, Nought, Newguise and Nowadays, may be called minor Vices, they correspond to some extent to the Deadly Sins, they act independently: they contend with Mercy and scoff at him, — 152. They are satirical against the Church, 134, 137. They use obscene and coarse expressions in attempting to lead Mankind away from his work only to receive in turn a sound beating with a spade, — 135. They dance, 78, sing, 323, seq., and are quite witty. In common with the Vice, they are active in calling up Titivillus, they support the Vice in his Court of Mischief and also in the scene where Mankind is tempted to hang himself, — 797.

4. *Nature.* — The Vice-rôle is here divided between the Vice, on the one hand, the Seven Deadly Sins and four other figures, on the other. The activity of the Vice, Sensuality, consists in his claiming before the World and Reason the gardianship of Man and in securing the service of Pride for Man, — 922[I]. After Man has been led astray by the World and Pride, Sensuality conducts him to the inn, and later reports what happened there, 1113—1158[I]: he is silent, or absent, during the actual temptation-scenes, 567—655, 689—709, 923—1035[I]: it is only in the second temptation of Man that he plays the part of a real tempter, but is again absent during

the scene between Bodily Lusts, Pride and Man, 165—318'ᴵ. The minor Vices participate independently in the action of the piece, especially in the first part, Mundus brings Man first into temptation, — 588ᴵ, W. Aff. introduces the Man to the Vice, Sensuality, 710ᴵ, and Pride plays quite independently a good Vice-rôle; he seeks with the help of the Vice to ingratiate himself into the favor of Man, — 840ᴵ, and is especially hostile to the Good; he makes taunting remarks in the presence of Man concerning his association with Reason etc., and, with the help of Wordly Affection, provides a fashionable dress for Man, — 1105ᴵ. Bodily Lusts figures as a messenger and go-between, — 220, — 291ᴵᴵ, Man sets him to keep together „all my company", 634ᴵ, that is, the Seven Deadly Sins. In the second part the Deadly Sins enter, more or less well characterized, and make preparations for the attack on Reason. Particularly interesting in this scene is the passage in which Envy praises Wrath, — 763ᴵᴵ and makes a fool of Pride, — 892ᴵᴵ.

The unified Vice-rôle. — That the earlier Moralities should contain so great a number of Vice-figures, each acting more or less independently, is due to the nature of the source from which their material was drawn; in the religious and didactic poetry in epic form, it was a very easy matter to handle a large number of allegorical persons, but on the stage, the difficulty of characterizing and distinguishing in the same piece between a number of persons of the same general type is almost insurmountable, as in *Castle of Perseverance, Mary Magdalene, Mankind* and *Nature.* Sins, human weaknesses, vices, are in the last analysis abstractions from the same idea, and, for the sake of unity, can be welded into one figure, namely, the Vice. That such a process of combination and elimination took place is at once made clear by an inspection of Table IV. This unification was completed at the end of the fifteenth century and is the basis of the distinction between the first and second group of Moralities.

The relationship of the Vice to the other evil powers is testified to very clearly by the Moralities themselves. A table of the genealogy of the Vice can indeed with considerable accuracy be made out. The Deadly Sins, for example, are called in *Castle of Perseverance,* 1400, the children of the devil. Again the personified Vice, Folly, in the *World and the Child,* 1506, is expressly designated as a summation of the Seven Deadly Sins, as the following passage proves. Conscience in instructing Manhood says:

„Sir, keep you in charity,
And from all evil company,
For doubt of Folly doing“.

Manhood: „Folly! what thing callest thou Folly?“
Conscience: „Sir, it is Pride, Wrath and Envy,
Sloth, Covetise and Gluttony,
Lechery the seventh is,
These Seven Sins I call Folly“, p. 258.

He then warns him to beware of Folly and Shame, p. 259, whereupon the Vice enters; after a short conversation with Manhood, Manhood asks him his name; Folly answers,

„I wis, hight both Folly and Shame“.

Manhood: „Ah, ah, thou art he that Conscience
did blame, when he me taught“, p. 263.

Already in *Cursor Mundi,* 10109, the Deadly Sins have been designated collectively as „folies“.

A second stage in the process is found in the *Lusty Juventus,* 1550, and the *Like Will to Like,* 1568. Here the Vice is called the son of the Devil; in the *Conflict of Conscience,* 1563, an „imp“ of Satan: In *Mary Magdalene*[2], Infidelity is called the son of Satan and he describes himself as being „the serpent's seed“:

„Look in whose heart my father Satan does me sow,
There must all iniquity and vice needs grow“, B i i.

In *All for Money,* Sin is the son of Pleasure and grandson of Money and himself the father of Damnation; of himself he says: „No sin can be without me“, B i i i, and then enumerates the Seven Deadly Sins. Furthermore he explains his own nature to the minor Vices, „as either of you contain one sin particularly, even so I contain all sins generally“.

In the controversial dramas of the reformation, a peculiar origin is assigned to the Vice in that he is made the son of the pope, *King John,* p. 8, *King Darius,* 770.

The devil and the Vice. — The matter stands thus: the devil does not become the Vice nor is the Vice simply a devil, as has been so frequently assumed; we must not confuse things which are so clearly distinct. The devil and Vice are, indeed, related in so far as all evil in society originates with the devil, but, as dramatic figures, they are distinct. The devil is essentially a theological-mythological being; he is the antithesis of divinity and sanctity, the friend of hell: as a dramatic figure, he has remained throughout almost unchanged.

The Vice, on the other hand, is an ethical person, he is an allegorical representation of human weaknesses and vices, in short the summation of the Deadly Sins: he is the antithesis of piety and morality and is the friend of an unrestrained worldly life. As a dramatic figure, he possesses in consequence of his composite origin, great versatility: he can, at pleasure, assume the rôle of a tempter or of a particular phase of vice or of vice in general. The specific human character of the Vice is shown in the various human rôles which he plays: Mischief, in *Mankind,* represents himself temporarily as a farm laborer, later as a judge in his „Court of Mischief", Hickescorner is a sailor, Sedition, in *King John,* and Hypocrisy, in the *Conflict of Conscience* are ecclesiastics, Sensual Appetite, in the *Four Elements,* represents himself as a tinker, also as a courtier, Hypocrisy, in the *Lusty Juventus,* at first, as a butcher, Ambidexter, as a gentleman, a lawyer, a student, Infidelity, in the *Mary Magdalene*[2], is a gentleman and a Pharisee, Riot, in the *Interlude of Youth,* is a criminal and ruffian, etc., etc. This conception of the Vice, Sensuality, in *Nature,* as a necessary factor in the human make-up, is suggested by Professor Brandl in his remarks on *Nature, Quellen,* XLIII: cf. also Symonds Shakespeare's Predecessors, p. 150.

The names of the Vices are almost never the names of specific vices: but are more or less general expressions, as for example, Mischief, Sensuality, Folly, Riot, Iniquity etc. These expressions are not, however, the most general expressions; they can all be generalized into a yet higher expression, namely the word, vice: Latin, vitium. This last step in the process of generalization was first made by Lydgate in

the *Assembly of the Gods,* written in the early part of the
sixteenth century, about 1520, where he describes a certain
character as the bastard son and agent of Pluto, in complete
armor, riding a serpentine, fire-breathing steed and followed
by a huge retinue of vices of all sorts, the chiefs of whom
are the „seven chief captains", the Deadly Sins, Sensuality,
Folly and Temptation; the name of this character is „Vice",
602. But Lydgate's example has not been followed in the
Moralities. Strange to relate, „The Vice", as a designation
for this allegorical figure, is first found in the Morality, *Respublica,* 1553, although the word vice in its ordinary signification, is uncommonly frequent in the Moralities[1]).

The Vice-names. — The Vice has no special proper name,
but the various Vices are given various allegorical names which
we may call Vice-names. Repeated use of any given name
is not unusual; e. q. Folly (Stultitia) appears in the *Castle of
Perseverance,* in the *World and the Child* and in *Magnificence,*
Sensuality (Sensual Appetite), in *Nature,* in the *Mary
Magdalene*[1] and the *Four Elements,* Mischief, in *Mankind*
and *Magnificence,* Inclination, in the *Trial of Treasure* and
in *Tomas More,* Hypocrisy, in the *Lusty Juventus* and the
Conflict of Conscience. For the rest, the Vice-names, except
in the case of some secondary Vices, appear but once.

As has already been said, the Vice-names are allegorical.
Even the name Hickescorner (the stupid scoffer), Nichol
Newfangle, Newguise, Haphazard (chance), Ambidexter,
Courage (impudent boldness), are to be allegorically understood and have more or less a satirical colouring. The alle-
gorical-satirical character of the Vice-names establishes at
once and beyond question a fundamental distinction between
the Vice, on the one hand, and clowns and fools, on the other.
These latter, in contradistinction to the Vice, have ordinary
proper names or pet names, as Boy, Man, Mouse, etc.

„*The Vice*". — The Vice has some times a double name,
the first part of which is the usual allegorical designation, as
Hypocrisy, and the second, the added expression „The Vice",

---

[1]) Pollard's remark is certainly strange, namely, that the Vice has
no place in the Moralities, English Miracle Plays, LIII.

e. q. „Avarice, the Vice of the play". Regarding the occurrence of this term in the lists of the players, stage-directions and text, the following observations can be made.

a) In the early Comedies. 1. In Heywood's *Play of the Weather*, about 1533, the expression „the Vice" occurs for the first time, and that in the list of the players. 2. The same is true of *Jack Juggler*, about 1562; 3. In Heywood's *Play of Love* the expression occurs once in a stage-direction, „Here the Vice cometh in running", etc. Brandl, *Quellen*, p. 200; (The list of players in this piece is lacking).

b) In the Moralities. 1. *Respublica*, 1553, is the first Morality containing the expression „the Vice". In the list of the players the character is given as follows: „Avarice, alias, Policie the Vice of the Play". Otherwise in this play this character is called simply Avarice. The word „vice", however, occurs frequently in this play and is so used that it is not clear whether it is meant to be a person's name or an abstract noun. For example, Verity says to Respublica; „whom thowe chosest are vices to be refused,

— — — — — — — — —

Than he y^t was Policie — — —·

— — — is most stinking and filthie Avarice. — — —

(he) cloked eche of these vices with a vertuous namme", *Resp.*, V, 3, 32—40; Respublica to Avarice:

„The best of youe is a detestable vice,

And thow for thie parte arte mooste stinking Avarice".

Avarice replies,

„Jesu! when were youe wonte so foule moothed to bee,

To geve suche nieck names?"

Resp., V, 6, 49—53. 2. In *King John* Sedition, the Vice, occurs in the list of the players, but was added by Collier, the Editor. 3. In the *Trial of Treasure*, 1567, Inclination the Vice occurs in the list of the players, and three times in the stage-directions. In one stage-direction the first component of the name is lacking, for example, „Gape, and the Vice gape" p. 273. In two other stage-directions, simply the first component of the name is used, as „Enter Inclination", p. 287. 4. *In Like Will to Like*, 1568, Nichol Newfangle, the Vice, occurs in the list of the players and in one stage-direction, otherwise simply Nichol Newfangle. 5. In Wit and Wisdom, 1579, occurs once the stage-direction „enter Idleness, the Vice“, otherwise Idleness. 6. In Thomas More, 1590, which has a play within it, the expression „Inclination, the Vice“ occurs in the text. 7. In *Mary Magdalene*[2], 1567, Infidelity, the Vice occurs in the list of the players and once in a stage-direction at his first entrance, otherwise 'simply Infidelity. 8. In *All for Money*, 1578, the stage-direction at his first entrance reads thus: „Sin being the Vice“ etc. In one stage-direction occurs: „Sin, the Vice“ and in three others simply „the Vice“, otherwise, in the list of the players etc., simply, „Sin“. 9. In the *Tide tarrieth for no Man*, 1576, „Courage, the Vice“ occurs in the list of the players and in the stage-direction at his first entrance, otherwise simply Courage. 10. In the three Wit-plays, the character, Idleness, which is common to all three, is designated as Vice only in the latest, — the *Contract of Marriage between Wit and Wisdom*, 1579, Shakespeare Soc., 1846, p. 12.

c) In the tragedies. 1. In *King Cambyses*, 1561, „the Vice“ does not occur in the list of the players, but does occur in four stage-directions, and that without the first component, as „enter the Vice“, p. 176, „the Vice run away“, p. 185, otherwise simply „enter Ambidexter“, p. 186 etc. etc. 2. The same is true of *Appius and Virginia*, 1564, in the case with Haphazard. 3. In *King Darius*, 1565, „the Vice“ occurs but once, and that for the first time in the text, namely, in the *Prologue*, thus, „The Vice is entering at the door“ with this stage-direction following: „the Prologue goeth out and Iniquity cometh in“; Brandl, *Quellen*, p. 362. 4. In *Horestes*, 1567, the case is entirely different; here we have the expression „the Vice“ consistently used throughout, in the title, the list of the players, and rubric; in the text, however, the expression does not occur.

Origin of „the Vice“. — From the foregoing facts it is evident, first, that the expression „the Vice“, excepting in *Horestes*, is nowhere consistently used throughout in the list of the players, text or stage-directions, and cannot, therefore, be original with the authors, otherwise we should expect greater consistency in the use of the term; and, second, that the term occurs in comedies and is used of characters which are, in reality, not all vicious, and is consequently again to be regarded as not original. But the question still remains, how came the term to be applied to the buffoon in John Heywood's *Play of the Weather*? It has been maintained, for example, by Swoboda, *John Heywood als Dramatiker*, p. 60, that Heywood borrowed the character of his buffoon from the Moralities. Perhaps so, but certainly not the name, for as already shown, the name first occurs in a Morality twenty years later. Further, it is observed, first, that the term occurs irregularly, first 1533, then 1553, then again 1561, after 1561, more frequently, but yet irregularly; the question arises, why this irregularity? And second, that in the text itself of the plays in question, no reference whatever to the Vice as a character is to be found, except in the one passage, in the prologue to *King Darius*; therefore, we must certainly conclude that the authors had nothing to do with the expression. It were certainly to be expected, had the authors themselves given the name to this figure, which is really the chief figure of these plays, that greater consistency in the use of the term would be found and that references to „the Vice" as such would occur in the text. The ordinary names of the Vice, as Hypocrisy, Folly, etc. are often referred to and played upon, why do we find no such references to Vice? The Vice often says, for example, in giving his name, „I am Folly" and „I am Ambidexter", etc., but nowhere does he say „I am Vice". The first unmistakeable reference to the Vice as a character in a play occurs in Shakespeare: „The formal Vice, Iniquity", and again, „The old Vice . . . with dagger of lath", and in Ben Jonson's prologue to „the *Devil's an Ass*", where the word is punned upon: „this tract will ne'er admit of our Vice, because of yours."

In so far as the extant plays can warrant a conclusion, this much is certain, that the term „the Vice" is not original in any play before *Horestes*. An hypothesis which admirably explains all the phenomena of the use of the term as set forth above, is, that it is the invention of the actors. The Moralities were, during the latter half of the sixteenth century, frequently acted: see for example the list in the repertoire of

5*

the company in the play of *Thomas More*. In all these plays there occurs a character which is in reality always one and the same, and that the chief character, but under various names: Folly, Hypocrisy, Iniquity etc. This character has been named by the actors as a matter of convenience „the Vice“, and by them the term has been inserted here and there in various plays; that is to say, the actors have done that which the authors have neglected, they have generalized the Vice-names.[1]) As this occurred in the period of deterioration of the Moralities, probably after 1560, at the time when the serious rôle of the Vice had fallen into the background and the farcical rôle was more and more on the increase, the term Vice came to be simply a synonym for buffoon. Hence the old definitions of the word: cf. Puttenham: „These Vices or buffoons in plays“; similarly, Cotgrave, under *mime*: „A Vice, a fool, a jester, scoffer“; and Dr. Samuel Johnson: „The fool or punchinello of the old shows“, and refers to Twelfth Night.

Earlier explanations of the word. — The occurrence of the word vice in Shakespeare has given rise to many explanations and notes; Douce is, however, undoubtedly correct in saying that the word „must be taken in its literal and common acceptation“, *Illustr. of Shakespeare*, I, 469.[2])

The function of the Vice and the Critics. — The earliest mention of the Vice as a special figure is in Stubb's *Anatomy*

---

[1]) In Sir David Lyndesay's Ane Satyre of the thrie Estaits in commendation of Vertew and Vituperation of Vice (first acted 1540?) Flattery, Falset and Dissait, three of the numerous Vice-figures enter, l. 601. seq.; they are first designated Vices in a stage-direction, p. 406, l. 838. Lyndesay uses the word vice only in the ordinary sense as the opposite of virtue.

[2]) Flögel, *Hofnarren*, 1789, p. 57, derives the word in the following manner; French vis d'âne < vis dase > vice = ass's head: similarly, Hanmer. Stevens derives the word from the French vis a mask, similarly, Brewer. Klein, *Geschichte des Dramas*, 13, p. 3, regards the word as the same as Latin *vice*, that is the vice is the devil's representative, „Des Teufels vice Teufel“. Theobald, *Shakespeare's Works*, Vol. V, 239, offers tentatively the following explanation: vice < O. E. jeck < GK. εἰχαῖ ι. ε. Fιχαῖ = Fελχ = „formal character, to put on the semblance of a better character, that is, to hide his cloven foot; he must put on a formal demeanour, and so moralize and prevaricate his words“. Warton explains the word as a derivation from device > vice = „a puppet moved by machinery“.

of Abuses, 1583; „For who will call him a wise man that playeth the part of a fool and a vice". Other early notices are in Puttenham's *Art of Poesie,* 1589, p. 97; Shakespeare in *Twelfth Night* and *Richard III,* Harsnett, 1603, Ben Jonson, 1610, Cotgrave, 1611. The famous passage in Harsnett's *„A declaration of egregious popish impostures . . . under the pretence of casting out devils"* etc. etc. 1603, is as follows:

> „And it was a pretty part in the old church plays, when the nimble Vice would skip up like a jack-on-apes into the devil's neck and ride the devil a course, and belabour him with his wooden dagger, till he made him roar, whereat the people would laugh to see the devil so vice-haunted".

We may also compare with this passage the song of the clown from *Twelfth Night:*

> „Like to the old Vice . . .
> . . . who with dagger of lath,
> In his rage and wrath,
> Cries aha to the devil".

It is very questionable whether the account by Harsnett [1]) should be regarded, as has been universally assumed, as applying to the Vice in general, for this passage describes new traits; the Moralities and the Tragedies give no indication whatever of any hostile relations between Vice and devil. Furthermore, Harsnett is here speaking of „Church plays", why is it that he does not use the then usual expressions „Moral Plays" or „Interludes?" The term „Church plays" can hardly be applied to Moralities. The above quoted passage from Harsnett must refer either to some lost Morality in which the Vice maltreats the devil, — but which, let it be noticed, must have been, in this respect, a very decided exception to all the extant Moralities — or to „Punch and Judy", which

---

[1]) Samuel Harsnett, 1561—1631, was Archbishop of York; he was a bitter polemist especially against the Roman Catholic Church, and against the conjuration of devils. Besides the above mentioned work, he wrote a similar one entitled *„A discovery of the fraudulent practices of John Darrel . . . detecting . . . the deceitful trade in the latter days, of casting out Devils,* 1599.

was common in England on feastdays and was, perhaps for this reason, regarded by Harsnett as a „Church play". It is not improbable that a late Mystery, which was customary to be played at York or in that neighbourhood, may have contained such a Punch-scene, as that referred to by Harsnett. The earlier history of Punch is rather obscure and requires an investigation.

In Punch and Judy we find a furious character carrying on a bitter feud with the devil. That Harsnett, as well as Shakespeare, should designate this figure as a Vice need not surprise us. They simply followed the fashion of the times; their use of the word simply indicates to what extent this character had deteriorated in the latter part of the sixteenth century. This interpretation of the Punch-Vice of Shakespeare has long since been hinted at by Jonson in Malone's edition of Shakespeare'sWorks in a note to „The roaring devil in the play" in *Henry,* V; he says, „in modern puppet-shows, which seem to be copied from the old farces, Punch sometimes fights the devil, and always overcomes him: I suppose the Vice of the old farce, to whom Punch succeedes, used to fight the devil with a wooden dagger", Shakespeare's Works, V, 566, Note 1; similarly again in a Note to *Twelfth Night,* IV, 95, Note 8. Compare also Dr. Samuel Johnson's definition already given, who, as do Shakespeare and Harsnett, appears to identify Punch and the Vice.

These old explanations of the functions and character of the Vice, are to this extent deficient, that they tend to represent the Vice simply as a buffoon. This, however, may be due to the fact, that during the second half of the sixteenth century, both the allegorical figure of the Good, the Vice's opponent, and that of Humanity, as the object of strife between the Good and the Evil, disappear from the stage. The real function of the Vice, as the opponent of the Good and the corrupter of humanity, i. e. the serious trait of his character, must also fall into disuse and naturally be forgotten. Thus there remain only the farcical traits of his character, and his name and title come then to be applied to the whole category of buffoons including Punch, but that Punch is a successor of the Vice or that Punch is to be identified with the Vice, follows by no means from this fact. Such an hypothesis would be even as great an error as to identify the Vice and the devil. The opinion of Roskoff, *History of the Devil*, I, 386, that the fool was evolved from the devil, and the clown („deutscher Hanswurst“) from the fool, is a purely fanciful statement, for which no proofs are furnished. The formula, devil became Vice and Vice became clown, does not apply to the facts in the case.

Harsnetts's works were certainly widely read; according to Theobald the „Discovery“ was Shakespeare's source for the names of fiends in *King Lear*. This work was certainly used by Jonson: cf. „Did you ne'er read Sir, little Darrel's tricks?“, The Devil's an Ass, V, 3; and it is not impossible that „the wooden dagger“ really owes its origin to the „*Declaration*“, although a reference to a wooden dagger occurs much earlier in *Like Will to Like*. It is really unfortunate that the entire learned world has so misunderstood Harsnett's account of the Vice and, without further examining into the matter, has up to the present time, persisted in applying his account to the Vice in general. All the authorities follow Harsnett almost verbatim, and vary in their accounts from one another hardly at all[1]). Some quote the passage in full, others refer to Harsnett, still others paraphrase or translate Harsnett without citing the source.

The Nature of the Vice-rôle. — The Vice-rôle is most intimately and vitally connected with the nature and structure

---

[1]) Theobald, (1767), *Works of Shakespeare*, V, 239.
Malone, (1790), *Works of Shakespeare*, I, pt. II, 20.
Douce, (1807), *Illus. of Shakespeare*, II, 305.
Sharp, (1828), *Dissertation*, p. 58.
Gifford, (1816), *Works of Ben Jonson*, II, 214.
Collier, (1831), *Hist. of Dram. Poetry*, II, 265, 270.
Ten Brink, *Gesch. d. engl. Litt.*, II, 318.
Klein, *Gesch. des Dramas*, XIII, 3.
Wright, *Hist. of Caricature and Grotesque*, 283.
Ward, *Hist. of the Engl. Drama*, I, 60.
Symonds, *Shakespeare's Predecessors*, 160.
Johnson's *Dictionary*.
Nare's *Glossary*.
The Century Dictionary.

of the Morality itself. The Morality consists essentially of three parts and has three principal persons: the first part consists of the convention between the Good and Man: the Man is generally good and industrious, and dutifully receives instruction and admonition. The second part is the Vice-rôle proper; the Vice, for the purpose of corrupting Man comes between Man and the Good. He slanders the Good, ingratiates himself with Man and entices him into a life of pleasure and sin, in which he also takes an active part. This work accomplished, he drops out of the play before the scene of the conversion of Man, — the third part. The formula for the Morality may be represented diagrammically thus, Good-Vice-Man, that is, the Vice is as an entering wedge between the Good and Man, and not as according to Brandl, Good-Man-Vice, that is, Man between the two powers of Good and Evil. As regards action, the Vice is the chief person of the Morality; all revolves about him as a centre of activity, in his unwearied efforts in causing mischief. His speeches and acts are from beginning to end seasoned with coarse humor and satire. The Vice-rôle is, accordingly, three-fold: first as the opponent of the Good, second as the corrupter of Man, third as the buffoon.

The Vice in various groups of plays. — Should the ground plan of a Morality vary from the plan above given, there generally follows a corresponding change in the Vice-rôle, e. q. *Hickescorner:* Here the temptation-motif is lacking from the simple fact that the representatives of Man are already corrupt. Further in *Nice Wanton,* a Morality constructed according to the French plan (Bien Avisé et Mal Avisé, and others): Here there are two representatives of Man, the one good and the other bad. The temptation-motif naturally is wanting.[1]

In the later Moralities and especially in the political Moralities, the rôle of the Vice suffers considerable modification. The reason for this is, that in these two groups of plays, excepting the *Trial of Treasure* and *Mary Magdalene*[2], the construction of the play is more or less incomplete, i. e. either the allegorical Good or Man is lacking. In the later

---

[1] Creizenach, *Gesch. des mod. Dramas,* I, 470, errs in saying that there is no English Morality constructed according to the French plan.

Moralities the figure of the Good practically disappears and the allegorical man is replaced by typical characters from various classes in real life, clowns and ruffians; this is particularly the case in *Like Will to Like, Tide* and *All for Money*. The part of the Vice in these plays consists in conducting the various bad characters to their fates, that is, he is a sort of nemesis; for example, Nichol Newfangle, while ringing the changes on the motto of the play, „like will to like“, brings Tom Collier and the devil together, and then, in the drinking-scene, Tom Tosspot, Philip Fleming and Hance. In a later scene he circumvents Cuthbert Cutpurse, Pierce Pickpurse and Ralph Roister and distributes to each, according to his deserts, a hangman's cord and a beggar's garb, as symbols of the natural results of a life of sin and dissipation. This didactic purpose of the play is essentially emphasized in the *prologue:*

> „Herein, as it were in a glass, see you may,
> The advancement of virtue, of vice the decay,
> To what ruin ruffians and roysters are brought, etc.“

In *All for Money* the Vice officiates at the court of All for Money, he ushers in the suitors, asks the necessary questions, and reports each case to the judge. Those whose claims have been rejected, he covers with jibes and taunts. In the *Tide tarrieth for no Man,* Courage rings the changes on „the Tide tarrieth for no Man“, and incites the various characters good and bad, Greediness, Courtier and the girl anxious to have a husband, to action.

In the political Moralities of the reformation period characters like *King John* or the widow England take the place of the Good and Man, but cannot be said accurately to correspond to them. In these plays the Vice is intentionally used for satirical purposes, to chastise the opposite political or church party or to lampoon evil conditious in the affairs of the Kingdom; the Vice-motifs, of hatred, slander, temptation, assume various new forms, or may be entirely lacking. The comic element is weak. The Vice, however, remains the principal character. In one play of this group, *King Darius*, there are two plots, side by side, one a Morality and one a Mystery, having really no relation to each other. The Vice in the Morality plot, strange to relate, has nothing whatever to do with the King and the chief persons in the Mystery-plot.

The Vice-Tragedies. — The rôle of the Vice is further modified in *King Darius, Appius and Virginia* and *Horestes.* Naturally in the tragedies the allegorical Man has no place whatever, the same is also true of the Good. The Vice here furnishes material for action, but is no longer the chief person in the plot. The mixture of the serious and the comical in the tragedies is, of course, not new, as we have already had it in the Moralities, and earlier still, in the Mysteries.

The function of the Vice in the Tragedies is two-fold; first, that of a tempter and deceiver, the objects of his deception being persons of high rank, kings, princes, judges; these he incites by means of fallacious arguments, to the commission of tyrannical acts. In regard to this function, the Vice may be regarded from two points of view, first: as having been borrowed from the Moralities, adapted, of course, to suit the changed conditions, or, secondly, he may be regarded simply as the personification of the evil that is in the hero, in accordance with the old idea which personified human faculties and passions. This view makes the Vice of the Tragedies to originate in the Tragedies themselves, subject perhaps to modifying influences of the Morality-Vice. The second function of the Vice is the buffoon of the play, in that he plays jokes, and fights with the clowns.[1]) Fischer, *Zur Kunstentwicklung der engl. Tragödie,* p. 34, has entirely misunderstood the Vice of the Tragedies. Ambidexter, in *King Cambyses,* he calls, e. q. „the first clown". To determine how nearly correct this is, one has only to compare Ambidexter with Hob and Lob and Snuff and Ruff of this play in order to perceive the mouth-piece of the poet, in that the poet makes the Vice to weep for the Queen, but Fischer here has overlooked the fact

---

[1]) The romantic Comedy, *Sir Clyomon and Sir Clamydes,* printed 1599, contains a character called Subtle Shift, in one place named the Vice. He puns on his name and serves by turn the two principal characters, misleads, however, no one, and, as far as the plot is concerned, is entirely subordinate.

that this weeping on the part of Ambidexter is pure burlesque, see below.

As examples of the Vice-rôle in later times, the black Ithimor and Mephistopheles in Marlowe, and Aaron and Iago in Shakespeare, have already in a general way been pointed out by Professor Brandl, *Quellen,* XCIV, but why not also add to these Edmund in *Lear,* Richard III, Don John in *Much Ado about Nothing,* and Antonio in the *Tempest?* These characters are all typical villains as well as those mentioned above. But until the Shakespearean villains have been satisfactorily investigated and classified, it seems idle to say whether these types are historically connected with the Vice or not. As possible forerunners of the villain-type are Lyon and Rewfyn in the *Conspiracy of the Jews,* in the Coventry Mysteries; Froward, servant of the Tortores in the *Buffeting Play* in the Townley Mysteries, and perhaps Brewbarret, (Strife-brewer) Cain's Servant, in the York Mysteries. The passage in the York plays containing this character is an interpolation of the middle of the sixteenth century: cf. Lucy T. Smith, York Plays, p. 37.

It is further to be remarked that the deed of revenge to which the Vice drives the hero in *Horestes,* although it may be, as Professor Brandl remarks, *Quellen,* XCIV, according to the conviction of the poet, a just act, is, nevertheless, thoroughly unnatural and cruel; indeed, Nature enters and attempts to dissuade him from it. In so far as this Vice here plays the rôle of a good counselor, he stands alone; the giving one's master good advice is not, as Professor Brandl would have us believe, a trait of the Vice, but of the trusty servant, an entirely distinct type: Compare, for example, Stephano in *Damon and Pythias,* written about the same time, the capital character, Will, in *Wit and Science,* 1570, and the fool in *King Lear.*

The Vice negatively considered. — Some negative considerations concerning the Vice are at this point appropriate and will serve further to define the character in question. 1. The Vice is not gluttonous. Eating and drinking play a very insignificant part in his actions; although much is said on these subjects, it is only to entice Man into the inn. Gluttony in

Nature, a minor Vice, appears, indeed, with a bottle and a cheese as his weapons, but this is exceptional. 2. Bodily exposure and the disgusting, in general, are notably absent, excepting perhaps in *Mankind,* 133 and 770, 773. 3. He never plays the husband. 4. The various social classes, such as laborers, peasants, the poor, the unfortunate, are never made the object of his satire and mockery, excepting monks, lawyers and the rich. 5. He is never stupid; he is full of conscious humor; he perverts and corrupts words and phrases, but always purposely and with satirical intent, but clownish misunderstanding is not characteristic of the Vice. Indeed, he sometimes betrays himself, for example, when his remarks have reference to himself, or when, as a result of bad habit, he swears at the wrong place. Furthermore, ineptitude and bungling in acting are unusual; the case of Ambidexter and *King Cambyses,* p. 234, is an exception. 6. The Vice has no peculiar or set mode of speech, verses with middle and endrhymes, which, according to Puttenham, *Art of Poesie,* p. 97, is a style particularly suited to the Vice, are to be found only in one song by Nichol Newfangle, *Like Will to Like,* p. 232, 233, and in the speeches of the parasite, Hardy Dardy, and of the three Vices, Pride, Adulation and Ambition in *Queen Hester.* 7. The Vice in the Moralities is never the servant of Man, except temporarily where he offers his service in order to win the good will of Man: for example, in the *World and the Child,* p. 263. In the tragedies; indeed, he assumes a quasi lackey position, in *King Cambyses,* p. 238, he figures as a messenger of the king, similarly in *Appius and Virginia,* p. 150, 151; in *Horestes,* at the close of the play he offers himself as a servant seeking a master, like the fool Cacurgus in Misogonous, Act. 4, sc. 3.

III. The Character of the Vice.

The Vice as a dramatic figure may be considered under two main heads, first: the character of the person as such and, second: as a dramatic figure simply. Under these heads may be collected and classified the various Vice-motifs, that is, all that the Vice does and says on the stage. (In this classification those motifs which the Vice has in common with other figures, such as oaths, obscenities, proverbs, phrases from foreign languages, and which are, therefore, not characteristic, are omitted). The character of the Vice is three-fold, according to his three-fold function: A. as an enemy of the Good, B. as the tempter of man and C. as a comical person.

A. The Vice as enemy.

The object of the Vice's persistent hatred is the allegorical representation of the Good.[1]) This character, the Vice's counterpart, named Reason, Pity, Mercy, Conscience, is an honorable exalted personage, dressed usually in a long cloak and wearing a long flowing beard. The animosity of the Vice towards this person finds expression in slander, mockery, threats and abuse of all sorts, including bodily assaults. Significant of this marked trait of the Vice is the meaning of some of the Vice-names, as Detractio, i. e. backbiter, and Hickscorner, i. e. scoffer. Since the Vice first gains an influence over man by destroying the influence of Reason, Pity or Mercy, it is clear that these two motifs, hostility on the one hand and temptation on the other, are intimately connected.

Other objects of the Vice's enmity, besides Reason. Mercy and the like, are the church and various social institutions and customs, and moral ideals, especially temperance, industry, piety. None of these are spared the lash of his satire. It is, however, to be noted that the Vice does not act from motives of revenge or the wish to populate hell; his attacks appear never to have ulterior aims.

Hostilities to the Good: Malice.

a) *Mockery:*

1. *Of his words* — [Man., F. El., K. J., K. D.].
Mischief interrupts the highflown speech of Mercy:
„Ye are full of predycacyone", Man., 47.
Similarly, Sensual Appetite with Studious Desire:
„Hast thou done thy babbling". F. El. p. 19.

---

[1]) For want of a better, handy designation for the opponent of the Vice and the friend of man, the expression „the Good" is used.

Iniquity says of Equity because of his long speech:
"He hath so many wordes in store". K. D., 470.
Mischief mocks the exalted style of Mercy:
"Ye are all to gloryende in yowur termys", Man., 760.
Similarly also Newguise:
"Yowur body ys full of englysh laten". Man., 121,
i. e. latinized words.
Sedition to King John in relation to England:
"For they are not worth the shaking of a peartree,
When the pears are gone; they are but dibble dabble,
I marvel how ye can abide such bibble babble". K. J., p. 7.
2. *Of his teachings.* — [Nat., F. El., W. C., L. J.].
Mundus designates them as "foly" Nat., 635^I,
Sensuality, as "folyshe counsell", Nat., 72^{II}.
Sensual Appetite, as "foolish cunning", F. El., p. 43, as
"sooth saws", F. El., p. 20.
Folly does not value the teachings of Conscience very
highly:
"He cannot else but preach . . .
I would not give a straw for his teaching". W. C., p. 264.
Hypocrisy to Juventus, who is on his way to hear a
sermon:
"Tush ... He will say that God is a good man".
[L. J., p. 73.
3. *Of his person.* — [Man., Nat., F. El., A. V., K. D., M. M^2.].
Mischief despises Mercy's store of wisdom:
"Yowur wytt ys lytyll, yowur hede ys mekyll", Man., 47,
also: "Yowur leude wndyrstondynge", 58.
Similarly Mundus with Reason:
"For he can neither good nor evil,
Therefore he ys taken but for a dryuyll", Nat., 637^I.
Nowadays taunts Mercy on account of his asceticism:
"Ye make no sporte", Man., 257.
Iniquity taunts Equity on account of his piousness:
"Whoo! have we more blessed scome to ye towne?" K. D., 291.
"This fellow is to good for mee", K. D., 468, and:
"Such another godsone", K. D., 499.
Sensual Appetite slanders the reputation of Studious
Desire:

„I promise you he hath a shrewd smell . . .
He savoreth like a knave", F. El., p. 20.
Similarly Iniquity to Charity.
„Ha, knave . . .
Thou lookest like ancient father and a old . . .
Tell mee one thynge, how doeth thy mynion?" K. D., — 54.
Infidelity to Knowledge of Sin:
„The devil is not so ill favored
Corrupt, rotten stinking and ill favored", M. M². , E. iii.
Haphazard says mockingly to Appius:
„Conscience was careless and sailing by seas
Was drowned in a basket and had a disease
For, being hard-hearted, was turned to a stone".
„For gifts they are given where judgment is none,
Thus judgment and justice a wrong way hath gone",
[A. V., p. 129.

4. *Opprobrious or taunting epithets.*
α) *In his presence.* — [Man., H., F. El., Y., T. T., K. D., M. M.²].
„Worshipful clerk", Man., 122.
„Jentyll Jaffrey", Man., 151.
„Gentle Harry", K. D., 1094.
„Yowur name ys do lytyll", Man., 251.
„My prepotent father", Man., 759.
„This churl Pity", H., p. 169.
„This caitiff", H., p. 171.
„This fellow", H., p. 172, K. D., 468.
Sensual Appetite says to Studious Desire:
„Now good even, fool, good even,
It is even thee, knave, that I mean", F. El., p. 19.
„Knave", F. El., p. 22, 36.
„Horeson knave", K. D., 124.
„Jackdow", F. El., p. 20.
„Whoreson", F. El., p. 36, Y., p. 26.
„Good John — a — peepo", Y., p. 25.
„Sir John", Y., p. 25.
„Master Charity", Y., p. 27.
„Ill favored lout", T. T., p. 294.
„Master Just", "Goodman Just", T. T., p. 278.
„Goodman Hobal", T. T., p. 277.

„Brother Snaps", T. T., p. 278, (with reference to the bridle).

„This gentleman", K. D., 110.

„Piss burned Cuckold", K. D., 418.

„Tom Narrownose", K. D., 848.

„Peter Turneup", K. D., 915.

„Nyck Candlestycke", K. D., 927.

„John Puddingmaker", K. D., 935.

„Thou pouchmouth knave", K. D., 973.

„False harlot" (Christ), M. M.², Fii.

„Beggarly fellow" (Christ), M. M.², Gii.

„Beggarly fool" (Christ), M. M.², Gii.

„Thief" (Christ), M. M.², Hii.

β) *In his absence.* — [Nat., W. C., F. El., T. T., K. D.].

„He is but a boy", Nat., 658ᴵ.

„Knave", Nat., 979ᴵ.

„These knaves", F. El., p. 43.

„daw", Nat., 1011ᴵ, W. C., p. 264.

„a straw", Nat., 1012ᴵ,

„a dryuyll", Nat., 637ᴵ.

„That bitched Conscience", W. C., p. 264.

„lousy lout", T. T., p. 295, (at parting).

„Master Just with his cankered courage,

What and old doting Sapience", T. T., p. 277 (at parting).

„Pyseburnd knave", K. D., 250.

„Shytten knave", K. D., 547.

„Peter Blowbowle", K. D., 174.

„Thys olde heretyke", K. D., 473.

5. *Nonsensical tasks and problems.* — [Man., F. El., K. D.].
Mischief proposes a nonsensical problem to Mercy:

„But, sir, I pray this question to clarify,

Dryff, draff, mysse masche —

Sume was corne and sume was chaffe", Man., 48—50.

Nowadays does the same:

„Now opyne yowur sachell with Latin wordes

And sey me þis in clerycall manere" etc., Man. 125.

Sensual Appetite, in order to entrap Experience, desires that he spell the word „Tom Cooper", F. El., p. 36. Nought imposes an obscene penance upon Mercy. Man., — 137. Iniquity proposes as an appropriate office for Equity that „He shall go play with my mother's pussy cat", K. D., 304; but Partiality is of the opinion that that were far too agreeable, he should rather be made to sell puddings, 308.

b) *Hatred and slander.*

1. *The Vice finds the person of his opponent uncongenial.* — [Nat., Man., H., F. El., T. T., K. D., M. M.[2]].

Sensuality says to Reason, while contending with him about Man:

„Forsoth, I trow about neyther we be good felowys",
[Nat., 311[1].

Sensual Appetite to Humanity:

„Though I do him (Studious Desire) despise", F. El., p. 20.

Sensual Appetite to the audience:

„For, by the mass, I love him not;
We two can never agree", F. El., p. 21.

Sensual Appetite to Studious Desire:

„Avaunt, knave, I thee defy", F. El., p. 22.

As Studious Desire comes into the inn, Sensual Appetite says to him:

„What art thou here! I see well, I,
The more knaves the worse company", F. El., p. 35,

the same is true of Inclination and Sapience, T. T., p. 277. Hickescorner makes peace between Freewill and Imagination so, that they all three may attack Pity. H., p. 169.

2. *The Vice seeks to drive his opponent away.*

The Vices wish that Mercy would go away:

„þe sonner þe leuer", Man., 250.

Inclination to Visitation:

„Will ye be packing", T. T., p. 294.

Iniquity to Equity:

„Go git thee home and talke with thy wenche", K. D., 420.

The *King Darius* in fact fairly swarms with such expressions:

„get thee away", 124,

„If thou go not hence to thee it will be death", 126.

Infidelity to Christ:

„It is best for you out of this coast to walk", M. M.², Fiii.

Sensuality rejoices as Innocence goes away, Nat., 656ᴵ, similarly again, Nat., 66ᴵᴵ, K. D., 170, 224 etc.

Iniquity takes measures to guard against being further molested by the Good:

„I must myself bestir
In my wrath and ire,
That they shall come no more" etc., K. D., — 561.

c) *Curses: imprecations.* — [Nat., W. C., H., F. El., T. T., M. M.², K. D.].

Sensuality:

„Let him go tho the deuyll of hell",
(i. e. Innocence), Nat., 657ᴵ.

Worldly Affection:

„Reason! Mary, let him go play
To the deuyll of hell", Nat., 1339ᴵ.

Folly to Manhood:

„A cuckoo for Conscience", W. C., p. 264.

Hickescorner:

„Yet had I liever see him (Pity) hanged by the chin",
[H., p. 171.

Infidelity wishes the same of Christ, M. M.², Gii.

Sensual Appetite to the audience:

„The devil pull off his skin:
I would he were hanged by the throat", F. El., p. 21.

Sensual Appetite:

„I beshrew thy father's son", F. El., p. 20.

Inclination:

„May the devil go with you and his dun dame", T. T.,
[p. 279.

„Farewell, in the devil's name", T. T., p. 295.

Infidelity:

„A poison take thee", M. M.², Fii.

Iniquity:

„I would you were in your graves", K. D., 1076.

d) *Threats: abuse.* — [Man., Nat., W. C., H., F. El., Y., T. T.].

Newguise to Mercy:
  „My brother wyll make you to prawnce", Man., 88.

Nowadays to Mercy:
  „Beware, ye may soon have a buffet", Man., 106 and
                                        [„trefett", 110.

Nought to Mercy:
  „If ye say þat I lye, I xall make yow to slyther",
                                        [Man., 109.

Sensuality to Reason:
  „Thou shalt auoyd myche sonar than thou wenuyst",
                                        [Nat., 267[1].

Pride:
  „I shall give him a lift", Nat., 851[1].

Folly:
  „Had I that bitched Conscience in this place,
  I should beat him with my staff,
  That all his stones should stink", W. C., p. 264.

Hickescorner: „This churl Pity
  Shall curse the time that ever he came to land",
                                        [H., p. 169,
  „And thou make too much I shall break thy brow",
                                        [H., p. 169,
  „With this dagger thou shalt have a clout", H., p. 171.

Sensual Appetite:
  „I shall make yonder knaves twain,
  To repent and be sorry", F. El., p. 37.

Infidelity:
  „We will rid this knave hence,
  Or else of his life I will soon make him weary", M. M.[2], Fi.

Iniquity to Equity:
  „Get thee away or I will thee slay", K. D., 50, 75, etc. etc.
  „Or with my dagger I will thee slay", K. D., 105,
similarly also T. T., p. 278, etc.

Riot:
  „I will lay him the visage", Y., p. 16; M. M.[2], Fii.

Riot:
  „Have on the ear
  And that a good knock", Y., p. 32.

Hickescorner:

„We will lead him straight to Newgate", H., p. 171.

Riot:

„We shall set him in the blocks" (stocks), H., p. 16, also p. 25. Pity is actually put into the stocks, H., p. 172, likewise, Charity, Y., p. 26, 27.

In *King John* the hostility of the Vice takes a peculiarly ecclesiastical turn. In this play one hears a great deal about excommunication and interdict. Armed with full power from Rome, Sedition covers the King and England with imprecations:

„Hold your peace, ye whore, or else ...
I shall cause the pope to curse thee as black as
[a crow", p. 4, also p. 66.
„Out with the popes bulls and curse him down to hell",
[p. 26. Compare further, p. 10, 25, 37, 38, 74.
„I am Sedition plain ...
Having you princes in scorn, hate and disdain", p. 18.

Dissimulation participates also in this hatred:

„A Johanne Rege iniquo, libera nos, domine", p. 25.

Since England is a woman, Sedition reveals his foul thoughts:

„What you i i alone! I will tell tales ...
And say that I saw you fall to lechery", p. 3.

Note. — Hostility towards Humanity. — [Y., N. W., K. C., M. M.[2]].

The Vice is in the rule friendly towards man. It is only when the Vice has failed in his designs to demoralize man, that he turns against him; herein he shows the meanness of his character.

Iniquity (aside), as Ismael is being led to execution: „Hang him" etc., „Let me be hang man" etc. N. W., p. 178. The Vice fears in this case Ismael's testimony against him. Infidelity is especially villainous in his treatment of the repentant Mary Magdalene; she creeps humbly to the Savior, Infidelity remarks complainingly:

„A sinner, quod he, yea, she is a wicked sinner,
A harlot she is" etc., M. M.[2]

Ambidexter addresses the dead King:
„How now noble King . . .
The devil take me, if for him
I make any moan", K. C., p. 245.

Riot imprecates Youth who has turned his back upon him:
„Fie on thee, caitiff, fie", Y., p. 38.

e) *Satire is the Vice's most strongly marked trait.*

1. *Against the institutions and usages of the church.* —
[Man., Nat., W. C., F. El., L. J., K. J., M. M.², M.].

α) *Monks.* — Newguise, perverting the words of the
Bible and at the same time quoting the devil: „Ecce quam
bonum . . . quod þe deull to the frerys — Habitare fratres in
vino" (instead of unione), Man., — 315, the same, in the perversion of the ordinal name: „Of the demonycall fryary" (instead
of dominical), Man., 144.

Sensuality in the joke at the expense of the cloistral life:
„She (Margery) hath entred into a relygyous place", meaning
a house of ill repute, Nat., 119².

Infidelity explains why his eyes are so crooked: „Like
obstinate friars I temper my look, which hath one eye on a
wench and an other on a book", M. M.², Ci.

Sin says, Sir Lawrence is no doctor except in the science
of „ducking women", further, that Sir Lawrence knows
neither Latin, Greek nor Hebrew, but can admirably „discharge
oaths" and can read playing cards to perfection, Diii, that he
can also drink, and, if need be, deal with the „Potter" — a
reference, evidently, to the Friar Tuck of the Robin Hood
Ballads. Money, Ei.

Folly pretends to have dwelt for years among monks and,
indeed, to have been crowned their king, W. C., p. 263.

β) *Indulgences and papal avarice.* — Nought: „Here
ys a pardone bely mett (be-limit i. e. belly-measuring, as explained by Professor Brandl),

„Yt ys granted of pope pokett", Man., — 135. He then
prescribes an obscene penance, — 137.

γ) The clergy and the church in general. —

Sensuality: „(Covetise) dwelled with a prest, as I heard say, For he loveth well

Men of the chyrche, and they him also", Nat., 999[II].

Infidelity likewise remains true to the church:

„The bishops, priest and pharisees do me so retain", M.
[M.², Ci.

Courage accuses the preachers of a lack of charity, he says to Greedines: „Not a preacher of them all in thy need will uphold thee, Try them who will, their devotion is small". Tide, Fiiii. The Vices ridicule the church music: they sing a mock anthem: „It is wretyne with a coll" etc., Man., 324, 328.

Similarly also Ignorance: „Give me a spade" etc., F. El., p. 48; he calls it a „peevish prick-eared song".

Sensual Appetite imitates mockingly the manner of the priests, probably, in this case, suiting action to words:

„Benedicite! I grant thee this pardon

And give thee absolution

For your sooth saws" etc., F. El., p. 19.

Similarly Iniquity to Equity: „Gods blessing, my son, I do thee give", K. D., 501, and to Charity: „Farewell, gentle Harry, I commit thee to God", 1095.

Avarice seeks a refuge under the protection of the clergy:

„Is then never a goode chaplaine in all this towne

That will, for awhile hide me under his gowne?"
[Res., 344—84.

Sedition describes the church orders and scourges the ignorance of the priests: „... Some to sing at the lecturn with long ears like an ass", K. J., p. 27.

Sedition describes the profits which the church is to derive from the interdict in England:

„Our holy father now may live at his pleasure,

And have habundance of wenches, wyues and
[treasure" ...

„Now shall we (the clergy) ruffle it in velvets, gold
[and silk,

With shaven crowns, side gowns and rochets white as
[milk", K. J., p. 65.

δ) *Saints.* — Sedition mentions St. Antonius' hog, K. J., p. 99.

Sedition, while being led to execution, says: „Pray to me whith candles, for I am a saint already", K. J., p. 99.

ε) *Relics: The catalogue-motif.* — Sedition exhibits his collection of relics, giving a complete inventory of all sorts of repulsive and impossible objects. The list begins with „a bone of the blessed trinity" etc., K. J., p. 47.

Hypocrisy describes likewise a long list of relics and sacred things of the Roman catholic church:

> „Holy cardinals, holy popes,
> Holy vestments, holy copes" etc., L. J., p. 65.

This effective satyrical method was first made use of in English by Chaucer, in the *Prologue to the Pardoners Tale*, and was borrowed from him by Skelton.

2. *Against social institutions and usages:* — [Nat., W. C., H., Y., T. T., L. W. L., Res., A. V., K. C., O., M. M.[2]].

α) *Women and marriage.* —

Sensuality describes Margery's life in the nunnery, i. e. in a house of ill repute:

> „Wedded, quod a, no, . . .
> They wyll not tary therefore,
> They can wed them selfe alone;
> Com kys me Johan, gramercy Jone;
> Thys wed they ever more,
> And it is the more to comend,
> For if a woman hap to offend,
> As it is theyre gyse", etc., Nat., 147[II].

Hickeseorner pretends to have traveled in „Land of Women, that few men doth find", H., p. 161—2.

Inclination pretends that love may be bought: „As for Venus . . . she is bought and sold always with treasure", T. T., p. 282.

Inclination hints that the women are immoral:

> „If ye chance to tell any tales of these gentlewomen,
> With flesh-hooks and nails you are like to be rent" etc.,

T. T., p. 287; likewise Haphazard, A. V., p. 124, and Courage, Tide, Dii.

Ambidexter scourges shrewish women:

„ . . . I have heard some say, —
That ever I was married now cursed be the day!
„Those be they that with cursed wives be matched,
That husband for hawks' meat of them is up snatched" etc.,

K. C., p. 232, likewise Haphazard and the Vice in O. and
refers to the story of Socrates and Xantippe. O., 1084—1114.
In M. M.[2] marriage is especially condemned; Infidelity to
Mary Magdalene: „For many incommodities truly be in
marriage", Cupidity agrees with him in this and Carnal
Concupiscence proposes „free love" as the proper moral
standard: „Take you now one, and then another". M. M.[2], Di.

Pride suggests to Youth, that it were really wise for him
to marry, but Riot, when he hears this, is of a very different
opinion:

A wife? nay . . .
The devil said he had liever burn all his life
Than once to take a wife", Y., p. 19.

The Vice is here, as he always is, the enemy of marriage, not
because the man by that means may be led to a moral life,
but because marriage is a sacrament of the church: cf. Mephistopheles, Marlowe's Dr. Faustus, 581.

β) *The legal profession.* — The law, as an object of
satire, is always brought into connection with the poor and
the oppressions, which they have to suffer.

Folly:

For I am a servant of the law,
Covetise is mine own fellow, — — —
And poor men that come from upland", etc., W. C., p. 262.

Similarly Sensuality, Nat., 999[II], and Res., V., 9, 32.

γ) *England and especially London.* — Folly, as well as
his ancestors, have from time immemorial lived in England,
he himself was born in London, W. C., p. 262; Hickescorner
relates that he sailed about the world in a ship called the
„Envy of London", H., 165. Among the passengers on this
same ship were all sorts of criminals, who had sworn always
to live in England, H., p. 164.

δ) *The rich.* — Against the rich but very little is said.
Inclination:

„The property of rich men undoubtedly he hath,
Which think with money to pacify God's wrath,
And health at their pleasure to buy and to sell".
[T. T., p. 295.

ε) *Against Fashions in dress the Vice has likewise surprisingly little to say.* In *Nature,* 748—781[1], Privy Counsel describes, just as Curiosity does in M. M.[1], his fine clothes; the colors are staring, the material rich and costly, the sleeves of his cloak are of themselves large enough to make a doublet and coat for some lad. His hair receives special attention day and night. To complete his out-fit he has a dagger and a sword so heavy that he requires a page to carry it.

Nichol Newfangle, whose Name, newfangle, means fond of what is new, relates what he, as an apprentice of Lucifer's, had learned to make: „gowns with long sleeves and wings";

„I learned to make ruffs like calves chitterlings,
And especially breeches as big as good barrels" etc.,
[L. W. L., p. 310.

In M. M.[2] the Vices are careful to instruct Mary Magdalene in the newest styles; first, her hair:

„In Summer time now and then to keep away the flies,
Let some of that fair hair hang in your eyes.
With a hot needle you shall learn it to crispe".

Pride:

„By your ears sometimes with pretty tusks and toys
You shall fold your hair like tomboys".

And when her hair at last begins to fade she must learn to dye it yellow:

„If the color of your hair beginneth to fade
A craft you must have that yellow it may be made",

and indeed „goldsmith's water" is peculiarly adapted to this purpose. Second, the use of cosmetics: Infidelity notices that Mary Magdalene has little pock-marks on her nose and that her complexion is too brown, but all this can be artificially remedied. Third, dress:

Pride naturally takes this matter upon himself:
„Upon your forehead you must wear a bon grace,
Which like a penthouse may come far over your face".
In front her dress must cut low, „That your white paps may
be seen", so that, as Cupidity suggests, young men may see
her white bosom and become incited to love. Carnal Con-
cupiscence adds that he has seen men actually bleed at the
nose at the mere sight of beauty. Further, according to In-
fidelity's advice she must straitly lace herself:
„Let your body be pent and together strained,
As hard as may be though you thereby be pained".
And finally besides cosmetics, laces, wires and the like, In-
fidelity recommends perfumery:
„Let your garments be sprinkled with rose-water,
Else your civet, pomander, musk . . .
That the odor of you a mile off a man may smell".

3. *Against morality in general.* — [H., F. El., L. J., T. T.].
Hickescorner reports how a large number of good pious
people were drowned in „the Rase of Ireland", H., p. 164.
Sensual Appetite would rather be dead than
„To pray, to study, or be pope holy", F. El., p. 20.
Riot thinks that the new learning on the part of the
reformers is a perversion of the natural order of things:
„Wilt thou set men to school
When they be old? . . .
Now every boy will be a teacher,
The father a fool and the child a preacher,
This is a pretty gear", L. J., p. 76.
He also scoffs at the puritanical custom of carrying Bibles about:
„At his girdle he hath such a book
That popish priests dare not on him look", L. J., p. 80.
Inclination (solus) discourses thus about truth and friendship:
„He that can flatter shall be well beloved;
But he that saith thus saith Christ,
Shall as an enemy be openly reproved.
Friendship consisteth now in adulation;
Speak fair and please the lust of thy lord" etc.
. . . „Behold how a lie can please some folks diet", etc.,
[T. T., 287, 288.

B. The Vice as the tempter and demoralizer of men.

Towards the person who represents man: Mankind, Manhood, Man, Humanity, Youth and the like, the Vice shows an entirely different phase of his being. The Vice appears as the embodiment of worldliness and sensuality, he is free from all restraints of religion and from all bonds of moral ideals. He is concerned only for one thing, that humanity shall give free rein to his inclinations, not however that a soul may be by this means ruined, but that man may be led to enjoy an existence of freedom and pleasure, the vicious ideal of happiness being in every sense the reverse of the spiritual.

a) *The Vice attempts to ingratiate himself with man who is at first unwilling and suspicious.* — [W. C., F. El., L. J.].

In these three plays humanity appears especially innocent, and, since he is just come from the instruction of the Good, he is particularly strict about all matters of piety. In W. C., p. 260, Folly greets Manhood in a jolly familiar manner. Being questioned concerning his occupation, he pretends among other things to be a great fighter and challenges Manhood to a bout. But as soon as Manhood learns that his name is Folly and Shame, he will have nothing more to do with him, the Vice begs to be taken into Manhood's service, and that simply for his keeping. As he finally consents to be called simply Folly, Manhood accepts him. It is interesting to note in this connection how slyly Folly manages; although the name of Conscience is frequently mentioned, Folly says nothing against him until he finds himself assured of Manhood's favor. In F. El., p. 20, 22 the matter of the temptation takes the form of a discussion. At first Humanity resents the slanderous attacks on Studious Desire: „Sir, he looketh like an honest man". The Vice continues to vent his opprobrium, but turns suddenly to Humanity with these words: „I am content, sir, with you to tarry" and „You cannot live without me" etc. Being asked, he explains his origin and function in the world thus: „I comfort the wits five", p. 21.

In L. J., p. 71, seq. Hypocrisy accosts Juventus and pretends to have known him in the past. Juventus at first does not recognize him. It is only after relating events from the early history of Juventus, that Hypocrisy succeeds in gaining a hearing: „Yes, I have known you ever since you were bore" etc., and „you and I many a time have been full merry". Being asked his name he gives an assumed name, Friendship, and goes then so far as to claim relationship with Juventus, who naturally enough is very glad to find an old friend again.

In M. M.[2] Infidelity in like manner claims old acquaintship:
> „I wis, mistress Mary, I had you in my arm,
> Before you were iii years of age". Bi.

He then officiously interests himself in her present affairs. Mary Magdalene is much worried about her dress, Infidelity blames the tailors. Mary is taken by surprise, whence all this knowledge about tailors and dressing? The motif, flattery is not uncommon with the Vice, but is peculiarly appropriate in this play, since, in this case, humanity is represented by a vain young woman. Infidelity offers advice, and suggests, as she is so beautiful and so rich and of such noble birth, that she should take special care to dress well and to live well.

b) *The Vice makes taunting remarks about Humanity's Manner of life.* — [Nat., W. C., F. El., L. J., Y.].

1. *On account of his dependence.*
Pride:
> „Me semeth ye saue not your honeste
> . . . a man of your behauyng
> Shuld haue alway suffycynt conyng
> Of worldly wyt and polycy . . .
> And not to be led by the ere,
> And beg wyt here and there
> Of every jak and pye", etc., Nat., 952—962[1],
> „I se well ye be but a very lad", 988[1];

Pride pretends to have thought Man at first worthy of some consideration, but now he repents having ever made his acquaintance, he becomes vexed and calls Man plainly a fool: „I wys ye ar but an ydeot", 1004[1].

2. *On account of his old fashioned dress.* —
Pride: „I fayth, I lyke not your aray", Nat., 1004[1].

3. *On account of his association with the Good.* —
Sensual Appetite is unable to understand how such a
person as Man can have anything to do with Studious
Desire:

> „Now, by my troth, I marvel greatly,
> That ever ye would use the company
> So much of such a knave", F. El., p. 22.

Similarly Sensuality:

> „Jesu, how may ye this life endure", Nat., 67^{II}.

4. *On account of his piety.*
Sensuality:

> „Where ys your lusty hart becom? . . .
> . . . I haue great maruell how ye may
> Lyue in suche mysery" etc., Nat., 73, 76^{II},

similarly, Riot:

> „He would have the a saint now,
> But a young saint, an old devil" etc., Y., p. 31,

and Pride:

> „He would . . .
> Make you holy ere ye be old" . . .
> „Thou wert a stark fool to leave mirth", etc., Y., p. 32, 33.

Hypocrisy says to Juventus, who is on his way early to
church with a prayer book under his arm:

> „A preaching, quod a? Ah! good little one
> By Christ, she will make you cry out of the winning" etc.,
> [L. J., p. 72.

Juventus is shocked at such words and attempts to defend
the Christian doctrine, Hypocrisy mocklingly replies:

> „Well said, master doctor, well said,
> By the mass, we must have you into the pulpit.
> ... Let me see your portous, gentle Sir John", L. J., p. 74.

W. Aff:

> „Why haue ye suche a spyced conscyence" . . .
> „I am sory and ashamed truely", Nat., 1050—1053^{I}.

5. *On account of supposed cowardice.* — Folly challenges
Manhood to fight, Manhood hesitates, being doubtful of the
Vice's ability to fight.

Folly:
  „No, sir, thou darest not, in good fay,
  For truly thou failest no(w) false heart", W. C., p. 261.

Riot roughly threatens Charity; Youth asks him to desist,
Riot turns immediately on Youth: „He turneth his tail, he
is afeard", Y., p. 26. Abh. Living to Juventus:
  „Who, you? nay ye are such a holy man,
  That to touch one ye dare not be bold;
  I think you would not kiss a young woman,
  If one would give you twenty pound in gold", L. J., p. 83.

  c) *The Vice leads Humanity into dissipation, after Humanity has surrendered himself to the worldly life.* — [Man.,
Nat., F. El., W. C., Y., T. T., M. M.[2]].

In *Mankind* Mischief opens his „Court of Mischief" and
makes Mankind take vows to steal, murder and the like,
Man., — 702. The usual method is for the Vice, in such
cases, simply to invite Humanity to go to the inn. Sensuality to Man: „. . . let us .ii. go
  To some tauern here bysyde", Nat., 1038[1].

What there took place, he relates afterwards, — 1144[1].
In F. El., p. 23, Sensual Appetite says to Humanity:
  „Well, then, will ye go with me
  To a tavern", etc., etc.,
Humanity agrees, the Vice thereupon calls the taverner and
orders the dinner, not without indulging with the taverner in
some coarse jokes; the taverner distinguishes himself credi-
tably in this encounter. It now occurs suddenly to the Vice
that more company is necessary, he proposes to bring in some
women of the town: „Then we will have Little Nell", etc.,
F. El., p. 26: again, p. 43, 44. Humanity does not appear to
be adverse.

  In Y., Riot says to Youth:
  „Youth, I pray thee have ado
  And to a tavern let us go,
  And we will drink divers wine,
  And the cost shall be mine . . .
  Yet thou shalt have a wench to kiss".
                                    [Y., p. 16: again, p. 23.

As in M. M.[1] so in M. M.[2], Mary Magdalene is led to the inn,
Infidelity:

„Will you resort with me unto Jerusalem,
A banquet they have prepared for you“, M. M.[2], Diii.

In W. C., p. 265, the invitation of the Vice assumes a
peculiar form in that the Vice takes occasion to express his
malice behind Manhood's back; Manhood proposes to drink
to his new acquaintance. Folly says to him: „Marry, master,
ye shall have in haste“, and turning to the audience says:

„Ah, sirs, let the cat wink,
For all wot not what I think;
I shall draw him such a draught of drink,
That Conscience he shall away cast“.

He then encourages Manhood to throw all restraints aside:

„Have, master, and drink well,
And let us revel, revel,
For . . . I would we were at the stews“.

He does the same later when he makes his exit:
Folly:

„Ah, ah! master, that is good cheer,
And ere it be passed half a year,
I shall thee shear right a lewd frere,
And hither again thee send“, p. 266.

In T. T., p. 272, the invitation to dissipation is represented allegorically; the vices are all present as persons, Inclination
formally introduces them to Humanity:

„Well, master Lust, first join you to me Inclination;
Next here with Sturdiness you must you acquaint;
Turn you about and embrace Elation;
And that wealth may increase without restraint,
Join you with Greedy-gut here in our presence“.

Lust is then suddenly seized with a violent cramp, Inclination explains this simply as a sign of his power over men.

The Vice, Mischief, suggests to Man in his desperation
that he hang himself and brings a rope and pole, Newguise
encourages him and shows him practically how it is done:

„Lo, Mankind, do as I do, þis ys the new gyse,
Gyff þe roppe just to thy neke þis ys my avyse“.

[Man. 791—92.

This suggestion to commit suicide is not, in this case, wholly original with the Vices. The thought had apparently already occurred to Man.

That Newguise really hanged himself at this time is clear from verses 795—97. Suicide as a mean of tempting men to destruction is ancient: cf. the old block-book, *Ars Moriendi* — first picture; there a devil is represented as calling the hopeless man's attention to one who had killed himself, the devil points also to the scroll „interficias te ipsum".

A desperation — scene occurs also in Skelton's *Magnificence*, 2312—52; Despair holds up the man's sins before him and intimates that faith, hope and mercy are now in vain and that the time for repentence is past. Mischief brings a halter and a knife: cf. the knife in *Ars Moriendi*. A similar scene occurs also in *Tide,* Gi, Courage holds up before Wastefulness his sins and Despair enters „in some ogly shape".

d) *The Vice stills man's reviving conscience.* — [W. C., L. J., Y., T. T., Confl., A. V., M. M.², Tide].

Manhood, just as he is about to begin reveling and drinking, expresses the fear that Conscience might yet find him;

Folly says:

„Tush . . . Conscience cometh no time here,

— — — — — — — — — — — — — —

For Knowledge have thou no care", W. C., p. 265.

Youth expresses his fears in the same way: „I would not that Charity should us meet", whereupon Riot threatens to give Charity a sound trouncing, Youth a second time becomes fearful just as he is going into the inn, Riot plies his victim thus:

„Let us go again betime,
That we may be at the wine,
Ere ever that he come".

Pride helps the matter on and declares himselt ready to pay the costs of the feast. Youth bravely seats himself at the table, but is suddenly seized with a peculiar fit; he says:

„Hark, sirs, how they fight". Thereupon R i o t, referring to the inner struggle between inclination and conscience, advises:

„Let not thy servants fight within thee,
We will go to the ale", Y., p. 23.

J u v e n t u s is fearful lest his friends find him in the society of Hypocrisy; Hypocrisy quiets him, teaching him how to play the hyprocrite thus:

„What are those fellows so curious

— — — — — — — — — —

Bid them pluck the beam out of their own eye,

— — — — — — — — — —

Call them papists, hipocrites" etc.,
„Let your book at your girdle be tied

— — — — — — — — —

And then will be said ... Yonder fellow hath
An excellent knowledge", L. J., p. 77.

H y p o c r i s y in Confl., p. 99 manages in the same way.

In T. T., p. 271, L u s t has fear of death and judgment, he had been reading Cicero and Paul, but such thoughts do not trouble him excessively; S t u r d i n e s s on the contrary suffers greatly on that score: „They cumber me pestilently". I n c l i n a t i o n proposes a remedy:

„Well, master Lust, such dumps to eschew . . .
. . . become a disciple of doctor Epicurus",

and further offers to bring in some jolly company, E l a t i o n and G r e e d y g u t.

In A. V., p. 128, J u d g e A p p i u s is halting between two opinions: „How am I divided" . . . „C o n s c i e n c e he pricketh me" etc., H a p h a z a r d quiets him, „tut man, these are but thoughts, C o n s c i e n c e has long since been drowned, you need concern yourself no more about him."

M a r y M a g d a l e n e is much concerned about her reputation and is doubtful about the propriety of free love, P r i d e undertakes to clothe the matter with a certain glamor, she should associate only with rich gentlemen who wear mantles with velvet collars, D i. After M a r y M a g d a l e n e had indulged in sin, she hears the „W o r d s o f t h e L a w" and becomes repentant:

> „O! Prudence, hear you not what the law doeth say,
> Exceedingly it pricketh my conscience", Eiii.

Infidelity makes an obscene joke and tries to lead Mary away: „Come away" and „Are you so mad him (the Law) to believe?"

> „These things are written to make folks afraid".

He argues further:

> „He speaketh of men, but no women at all,
> Women have no souls", Eiii.

As the Law goes out, Infidelity feels of her pulse and pronounces her sound, Mary says her body is sound but her conscience is very sick. Infidelity is then about to discuss this subject of conscience, when Christ enters; there is then nothing left for the Vice to do but to scold: „Do you love me?", „You have a wavering wit" etc. Similarly in *Tide,* Bii, Courage scolds Greediness, who had been listening to a sermon and as a result was complaining of remorse of conscience.

This motif — the stilling of the conscience of man, is a characteristic Vice-trait. It is also made use of in *Richard* III, I, 3, in the case of the two murderers, who resemble the Vice in many ways. The Second Murderer hears accidently the word conscience and is afraid, the other says mocking, what are you afraid! After a little while he asks again how it is with him; his pal confesses that he still feels some traces of conscience, but this is all finally put effectually to flight when he is reminded of the reward.

e) *Arguments.* — [Nat., W. C., F. El., Y., L. J., M. M.2.] The Vice resorts to the use of arguments to lead man astray, as Lucifer does in W. But, as a rule, this cannot be said to be a marked trait of the Vice. Serious Arguing does not agree very well with mockery and buffoonery, and it presupposes further an antagonism which, as between himself and man, the Vice seeks above all things to avoid. The Vices may, accordingly, from this point of view, be divided into two classes, those who are more serious and argumentative: type, Sensualiy in *Nature,* and those who do not argue at all: type, Mischief in *Mankind.*

1. *Religion and studiousness are to be rejected.*

Sensuality:

„Without ye take some other wayes,
By my throuth yt wyll shorten yonr dayes", Nat., 79[II].

Sensual Appetite:

„It will you bring
At last into your grave", F. El., p. 22.

Infidelity:

„Never attend you to law nor prophecy,
They were invented to make fools afraid" — —
„God? tush, when was God to any man seen?"
„Homo homini deus." M. M.[2] Cii.

Idleness in her song is especially opposed to study, W. Sc., p. 374.

2. *Our fathers were certainly in the right.* — This argument is especially resorted to in fighting the reformation.

Hypocrisy to Juventus:

„Was not your father as well-learned as ye?
And if he had said then as you have now done,
I-wis he had been like to make a burn". L. J., p. 74.
He repeats this argument, p. 75, 76.

3. *There is always time enough to repent.* —

Pride to Youth:

„I trow that he would
Make you holy, ere ye be old; — — —
It is time enough to be good,
When that ye be old". Y., p. 32;
„Thou art not certain of thy life;
Therefore thou wert a stark fool
To leave mirth and follow their school", p. 33.

Infidelity:

„You shall never be younger". M. M.[2], Bi.

4. *The Vice recommends himself.* —

Folly says of himself that he is everywhere highly respected, and therefore is not to be despised:

„For Folly is fellow with the world
And greatly beloved by many a lord". W. C., p. 264.

Sensual Appetite explains the doctrine of sensuality, his fundamental proposition is this: „Ye cannot live without me", F. El., p. 21, for the five senses are essential to life and the peculiar office of Sensual Appetite is this: „I comfort the wits five". Infidelity recommends himself to Mary Magdalene as a trusty counsellor: „You cannot trust a wiser", M. M.[2], Bi.

5. *Worldly prudence.* Suggestion maintains that Paul, Christ and David dissembled in order to save themselves from their enemies, Confl., p. 109, further, that we must be wise in the choice of evils and trust God, who is merciful rather than men, p. 111.

6. *The argumentation in the Tragedies differs from that of the Moralities in that it lies, as it were, partly in the names of the Vices.* Thus Ambidexter says to Sisamnes, it is very stupid of you to be so conscientious, no one will dare to impeach you, the opportunity is favorable, double dealing is all that is necessary: „Can you not play with both hands and turn with the wind?" and explains to him a plan whereby his brother's crown may be gained. K. C., p. 187, 188.

Haphazard proceeds in the same way; as he hears the complaints of the lovelonging Appius, he urges him to try his chances:

„Why, cease, sir knight, for why perhaps of you she shall
[be bedded" — —
There is no more ways, but hap or hap not" etc., p. 127;

Conscience and Justice appear, Appius is sore afraid, Haphazard resorts once more to his argument of chance:

„It is but in hazard and may come by hap;
Win her or lose her, try you the trap". A. V., p. 132.

He then proposes the wicked plan of kidnapping Virginia. In O. the Vice pretends to be a messenger of the gods, gives himself the name of Courage and then says to Horestes spurring him to action with those words: „Seke to dystroye, as doth the flaming fier", etc., O., 287.

f) *The Vice reminds man of former delights.* — [Nat., F. El., Confl.]

Schould the man in the course of the play repent, it becomes necessary for the Vice to retain control over his victim. Sensuality reminds Man of his old companions: „Meny a good felow wold make great mone", Nat., 81[II], and he goes even so far as to weep because Man has so shamefully neglected them, 82[II]. The weeping and the news of the old friends appear to have effect, the Vice seizes then the favorable moment to remind Man of his old flame, Margary. Sensual Appetite in F. El., p. 43, in a similar way, reminds Humanity of the happy times spent in tavern:

„Wether thought you it better cheer,
At the tavern where we were ere,
Or else to clatter with these knaves here". F. El., p. 43.

In Confl., Hypocrisy, who is to be regarded as an inquisitor, tries zealously to persuade Philologus to recant, first, he expresses great sympathy, promising him mercy if he will only yield; then he holds up before him the horrible consequences of such persistency: „Your zeal is too hot; which will not be quench, but with your heart blood". Confl., p. 87. Sensual Suggestion aids Hypocrisy in this matter, thus he relates in the hearing of Philologus how, as he came through the streets, he heard a certain woman, the mother of several helpless innocent children, wailing because of the persistent, stiffnecked conduct of her husband - this was of course the wife of the victim, Philologus, Confl., p. 93.

C. The Vice as a comical figure.

The purpose of humor in the serious dramas is not infrequently adverted to in the prologues and the admixture of the comical element explained and justified; since the common man listens very willingly to the comical, he may, therefore, be induced to hear the serious also, if it be interlarded with fun. This is clearly expressed in the prologue to the *Four Elements*:

„This philosophical work is mixed
With merry conceits to give men comfort

> And occasion to cause them to resort
> To hear this matter".

Cf. further, among others, the prologue to *Trial of Treasure* and *Like Will to Like*. Thus the authors introduce the comical intentionally and seek to justify this on utilitarian grounds.

The frequent expressions in these prologues: „mirth", „merry conceits", and the like, refer unmistakably to the Vice and his rôle as a comical person. Sometimes the Vice says of himself that he has come purposely to create fun. But this humor of the Vice is not without a tinge of the maliciousness, which is an essential part of his make-up, the Vice is not a purely humorous character. One feels that what he says and does has always a background of maliciousness and satire. The Vice's witticisms are:

a) *Playing upon-words.* — Both the number and the variety of the punnings in the earlier serious dramas are not great, especially in comparison with Shakespeare.

1. *Playing upon the sound chiefly.* — [Man., L. J., Y., K. J., L. W. L., Res., A. V., O., M.]

Nought invents the word „trefett", Man., 110 to match „bofett" (a blow), 106; „trefett" is perhaps a tripple blow. Nought makes a similar play with the word „shett", Man., 773, referring to the preceding „shott" as is evident by his remarks: „I am doyng my nedyngs", 770 „I haue fowll arayde my fote", 771.

Sedition will „with the pope hold so long as I have a hole in my breech", K. J., p. 4; again Sedition: „It were folly such louce ends for to lose", K. J., p. 74.

Hypocrisy: „breakfast" > „pie-feast". L. J., p. 78.

Avarice attempts to teach the stupid Adulation the name Reformation, thus: „Ye shall learn to (do)-sol-fe-re-for-ma-tion. Sing on now; re-for-" etc., Res., II, 4, 68.

Riot, bringing a chain: „Is not this a jolly ringing?" Y., p. 27. Referred to the preceding „a pair of rings", „ringing" acquires the significance to put in irons.

Tom to Nichol Newfangle: „Your presence hath made my heart light".

Nichol replies: „I will make it lighter", L. W. L., p. 347 referring to the hanging which awaits Tom.

Sin to Moneyless: „Thou art sure shortly to play sursum corda", Money, Di.; cord(a) = 1. the hangman's rope, 2. the hearts; Sursum = upward, i. e. hanged.

Haphazard: „He never learned his manners in Siville", A. V., p. 151; Seville = 1. a city in Spain, 2. civil, polite: cf. the Taverner's: course > coarse, F. El., p. 25.

Etymological punnings are rare. — Avarice, being ad- d.essed as „founder", answers snappishly, „Founder me no foundring", Res., I, 3, 50.

Haphazard plays upon his name thus: „Therefore hap and be happy", A. V., p. 151, so also,

„When he hazards in hope what hap will ensue?" — —
„A ploughman — — May hap be a gentleman" — —
— — „Hap may so hazard", etc., p. 124,
— — „but hap or hap not,
Either hap or else hapless" —· —, p. 127,
also p. 129, 130, 147, etc., cf. also:

Hempstringe:
„Hange me no hanginge", O., 372.

2. *Misinterpretations; the Vice is fond of distorting words by substituting syllables similar in sound to produce a satirical effect.* — [Man., K. J., Res., K. D., Confl., T. T., L. W. L., M.]

α) *Chiefly single words.* — Newguise: „bely mett" (be-limit), Man., 134: see above p. 110 β; „demonycall" (dominical), Man., 144; „in vino" (in unione), Man., 315; „Yowur neglygence" (your reverence), Man., 445.

Dissimulation to Ursurped Power:
Your horrible holiness", K. J., p. 34.

Sedition betrays himself thus:
„I have a great mind to be a lecherous man —
I would say a religious man", K. J., p. 12.

Likewise, Hypocrisy:
„I speak mischievously —
I would say in a mystery", Confl., p. 46;

„desolation — — — consolation“, Confl., p. 66;
„I will be the noddy —
I schould say the notary“, Confl., p. 79.

Iniquity attempts to claim relationship with Equity because of the similarity of their names, but in vain; he therefore remarks: „That preposition *in* is a pestilent fellow“, K. D., 838.

β) *Meaningless words and phrases, echoes.* —

Lucifer says to Nichol Newfangle:

— — — „that thou adjoin like to like alway“.

Nichol replies: „That I eat nothing but onions and leeks alway“, L. W. L., p. 312, 313.

Lucifer requires the Vice to salute him with the words: „All hail, o, noble prince of hell“, Nichol Newfangle twists this into the following: „All my dames cows tails fell into the well“ etc. etc., L. W. L., p. 316.

Dissimulation sings the litany:

„Sancte Dominice, ora pro nobis“,

Sedition overhears him, aside; „Sancte pyld monache, I beshrow vobis“. K. J., p. 25.

The following interesting stage-direction occurs in *All for Money*, Ciii, „Here the Vice shall turn the Proclamation to some contrary sense every time All for Money hath read it, and here followeth the proclamation“; but the Vice's misinterpretations, in this case, are not given. Here it is thus expressly provided that the Vice shall improvise — a custom on the part of clowns against which Hamlet later so vigorously protests. The Garcio, Cain's servant, twists the words of his master's proclamation in the same way, Townley Plays, *Mactatio Abel*, 418—438.

γ) *Malicious side-remarks, the Vice utters bitter truths, his words generally rhyming with those of his opponent; mimesis.*

Lust to his mistress Lady Treasure:

„My lady is amorous and full of favor“,

Inclination: „I may say that she hath an ill favored savor“.

Lust: „what sayest thou?"

Inclination: „I may say she hath a loving and gentle behavior". T. T., p. 292: Other examples, T. T., p. 288, 289, 291, Confl., p. 77, K. D., 266, *Marriage of W. and W.*, p. 19. Shakespeare makes use of this motif, Rich. III, III, 1: Gloucester makes a mysterious remark, the little prince asks what he means, Gloucester replies with words of a similar sound but different meaning explaining then to himself: „Thus like the formal Vice, Iniquity, I moralize two meanings in one word". The word „formal", here, is equivalent to dealing with expressions of similar sound, that is, equivocating.

3. *The Vice uses a single word in a double sense, and that but once.* — [F. El., L. J., Y., Res., T. T., L. W. L., A. V., M. M. 2.)

Sensual Appetite:

„Well hit, quoth Hyckman, when he smote
His wife on the buttocks with a beer bottle". F. El., p. 19.
Hit = 1. met, 2. struck.

Riot:

„The devil said he had liever burn all his life
Than once to take a wife". Y., p. 19.
Riot here refers to the doctrine of Paul, *I. Corinthians* VII, 9, and to the fire in hell.

Inclination:

„I must tune my pipes first of all by drinking", T. T.,
[p. 274.
Pipes = 1. musical instrument, 2. oesophagus. Tune = to put in order.

Nichol Newfangle presents to a spectator a playing card, namely, the knave, saying: „Stop, gentle knave, and take up your brother". L. W. L., p. 309. He also uses „hole" in two senses, p. 311.

Inclination:

„For having this minion lass,
You shall never want the society of Pallas". T. T., p. 282.
Want = 1. to lack, 2. to need.

Haphazard: „At hand, quoth pick-purse, here ready am I",
A. V., p. 129.

The ambiguous expressions, *double entendre* proper, are especially characteristic of Res.: Adulation: „We ... travaile for your wealth" etc., Res., III, 2, 7; Avarice: „And this is all yours", namely the money which he had embezzled, Res., V, 9, 107; other examples, II, 3, 29, IV, 4, 81, seq.

Similar expressions occur in M. M. 2; Carnal Concupiscence says to Mary Magdalene with reference to the action of Infidelity: „He hath been diligent about your cause". Also in a bad sense; Hypocrisy says to Abhominable Living: „Be good to men's flesh" etc., L. J., p. 85, Infidelity to Mary Magdalene:

„They had liefer have you naked —
— — Than with your best holy day garment". M. M. 2.

The style of Mary's dress had just been the subject of discussion.

4. *In dialogue; complex puns.* — [Man., F. El., Y., K. J., K. C., K. D., Res., T. T., M. M.², M.].

α) *The Vice purposely uses a word ambiguously, but not being understood, he is asked to explain his meaning; his answer completes the pun. Examples are infrequent.* Sensual Appetite wished to order from the Taverner small birds, they are „light of digestion", since they are „continually moving". As soon as the Taverner understands the joke, he rejoins, I know of a still lighter flesh, a woman's tongue, „for it is ever stirring". F. El., p. 25.

Sedition makes an obscene pun in the form of a riddle on the words hole, holy, K. J., p. 35. Likewise Newguise, on the word „marriage", but proceeds with the explanation without waiting for intermediary question, Man., 331.

β) *A player intentionally uses a word ambiguously, the Vice takes up the jest and developes it farther. Examples are not frequent.*

Riot describes his experiences as a thief and murderer; Youth opines:

„— — — thou didst enough there,
For to be made knight of the collar".

Riot:

> „Yea, sir I trust — — —
> At the next sessions to be dubbed a knight", Y., p. 15.

Collar = 1. Collar of an order, 2. Hangman's noose. Sessions = 1. Sessions of Parliament, 2. sessions of the court.

γ) A player uses a word in the ordinary sense; the Vice, however, gives the expression an other meaning based upon some association of ideas. Catching-up.

Mischief catches-up Mercy's words: „The corn xall be sauyde þe chaff xall be brente", Man., 43, and gives them a ridiculous interpretation; the word „corn" suggests „thresher", „bread" and „baking", on the one hand, and, on the other, „horse" (which eats straw), „fire" (straw being used for fuel) and „cold", the contrary of warmth, fire. In a similar manner Newguise catches-up Mercy's words: „Yf a man haue a hors and kepe him not to hye", etc., Man., 230, applying them to his family life: „I haue fede my wyff so well" etc., 235.

Sedition pretends to misunderstand the words of the king: „So thou powder it with wisdom" etc. (i. e. season your conversation with wisdom); Sedition replies: „I am no spicer" (dealer in spices). K. J., p. 3. Similarly: „fed with ... ceremonies" > „they eat both flawns and pygyn pies". K. J., p. 4.

Dissimulation says: „At last I have smelled them out". Sedition replies: „Thou mayst be a sow, if thou hast so good a snout". K. J., p. 30.

Iniquity catches-up Charity's words: „A fervent love we keep in store", he replies, yes, he would surely do that, namely, keep a tight hold on his money". K. D., 86.

Ambidexter gives an obscene meaning to the word „corner". K. C., p. 178, so also, Infidelity, to the expression „pricke of conscience", M. M.[2], Eiii.

Money asks Sin about his ancestry: „from what stock you are descended?", but Sin understands stocks: „The last stocks I was in was even at Bambury". Money Ciii. — Stock = family, race; stocks = an instrument of punishment.

Adulation to the other Vices: „Yea we must all hold and cleave together like burres.

Avarice remarks maliciously: „Yea, see ye three, hang and draw together like furres", Res., I, 3, 132.

Lust to Lady Treasure: „I love thee, in faith, out of measure".

Inclination (aside): „It is out of measure, indeed, as you say". T. T., p. 289.

b) *Naïve, self-evident answers.* — [Y., T. T., L. W. L., Res., K. D., O.]

On Riot's first appearance Youth asks him: „Who brought thee hitherto"? Riot replies naïvely and literally: „That did my legs". Y., p. 13. The same sort of witticism is used by Titivillus, Man., 439, and by the First Murderer, Richard III, I, 3.

Nichol Newfangle after being beaten by the rogues, asks: „tell me am I alive or am I dead?" L. W. L., p. 351.

Avarice asks Oppression, who seems to be weary: „Where have you lost your breath". Res., III, 5, 3.

The Vice, as he contemplates attacking the clowns, remarks prudently: „(It is) good sleepinge in a whole skynne", O., 103.

Iniquity: „Why, man, it is yellow", namely, the „yolke of an egge", K. D., 955; again: „Two dishes maketh a platter", 890.

Inclination: „To it, and I will either help or stand still". T. T., p. 292.

Nichol Newfangle: „An owl is a bird". L. W. L., p. 332, again: „knaves flesh is no pork", p. 332.

Courage: „Needs knaves you must go, for so you came hither". Tide, Diiii.

Profit, one of the minor Vices, replies in the same tone: „But here we found thee most knave of all and so we leave thee".

Idleness explains his name: „I am ipse, his even the same". W. W., p. 16.

c) *Nonsense.* — Nonsense, or lack of connection in discourse, is, according to Professor Child, a distinctive characteristic of the Vice. This variety of witticism consists generally of the combinations of irrelevant matters and furnishes the speeches of the Vice, that fantastic element, which corresponds to the grotesque in his costume.

1. *Alliterating and rhyming words.* — [Man., K. J., L. J.]
Mischief: „dryff, draff, mysse masche". Man., 49.

Sedition: „dibble, dabble, bibble babble". K. J., p. 7. With these words the Vice, means to imply that what his opponent says is of no consequence.

Haphazard, seeing the servants fighting: „What culling and lulling ... what tugging, what lugging, what pugging", A. V., 120.

Hypocrisy, as he sees Juventus kissing the wench, cries out jealously:

„What a hurly burly is here,
Smick, smack — — — You will go tick, tack", L. J., p. 85.

2. *Irrational speeches.* — [H., F. El., A. V., K. C., T. T., Tide, W. W.]

Hickescorner mentions among the many lands which he has visited:

„The land of Rumbelow
Three mile out of hell". H., p. 162.
Rumbelow is a fantastic word taken from the refrain of an old sailors' song: cf. Halliwell.

Similarly, Inclination weaves an absurd remark into his otherwise rational speech:

„I can remember since Noe's ship
Was builded on Salisbury Plain". T. T., p. 267.
Talking nonsense, in vaunting their exploits, is a marked characteristic of those Vices, who would play the Miles Gloriosus.

Sensual Appetite:
„I was at a shrewd fray — — — I have slain them
[every man,
Save them that ran away". F. El., p. 41.

Courage:
„Where good wife Gull broke her good man's pate
In came her man to make up the number

Who had his nose shod with the steel of a skumber.
But, in fine, these three began to agree,
And knit themselves up into one trinity — — —
For very love they did kill one another,
And they were buried, I do well remember,
In Strawnton's strawn hat vii mile from December.
They had not been dead the space of a day,
But four of those three were thence run away,
The Constable came — — —
And because they were gone he did them kill.
I was twice smitten to the ground,
I was very sore hurt but a had no wound", Tide, Bi.

Inclination wishes to give Lust and Sturdiness an impression of his own importance: „Look on this leg — — —
I can remember when it was no greater than a tree". T. T.,
p. 269: Query; how big is „a tree"?

Ambidexter, whose character reminds one of Thersites,
boasts of that which he is about to do:

„I am appointed to fight against a snail — — —
If I overcome him, then a butterfly takes his part,
His weapon must be a blue speckeled hen.
If I overcome him, I must fight with a fly,
And a black pudding the fly's weapon must be".

[K. C., p. 176.

Haphazard's account of himself, at his first entrance, contains a string of nonsense, extending through some 36 verses,
for the most part alliterative, he tells what he is:

„A lawyer, a student or else some country clown,
A louse or a louser, a leek or a lark" etc., A. V., p. 118.

Also p. 124: As peacocks sit perking by chance in a plumtree",
p. 130: „And geese shall crack mussels", and p. 134: Run for
a ridduck" etc.

In F. El., p. 49, the speech of Ignorance consists of 24
lines of absurdities. In W. W., p. 49, in the midst of much
that is nonsensical Idleness reports seriously the misfortunes
which have befallen him since the last scene.

The talking of nonsense is used to a very different purpose by Haphazard; in the midst of much nonsense, to which he gives expression, is an element of truth; thus, as in the confused speech of Hamlet, which this passage strongly resembles, Haphazard's speech follows immediately upon the evil determination of Appius, „I will have Virginia", and effectively presages the coming catastrophe, namely, the approaching execution of Appius, Claudius and Haphazard:

„I came from Caleco the same hour,
And Hap was hired to hackney in hempstrid:
In hazard he was of riding on beamstrid.
Then, crow-crop, tree-top, hoist up the sail,
Then groaned their necks by the weight of their tail:
Then did Carnifex put these three together,
Paid them their passport for clust'ring thither".

Appius answers in great surprise:

„Why, how now, Haphazard, of what dost thou speak?
Methinks in mad sort thy talk thou dost break" etc.

The expression „these three" has reference evidently to: 1. hempstrid (hangman's rope), 2. beamstrid (beam, or tree), 3. hoist up the sail (to hang).

3. *Irrelevancies.* — [Man., Nat., F. El., L. J., Y., K. J., K. C., A. V., K. D., L. W. L., T. T., M. M.[2], M., Tide.]

Related with the preceding are many expressions, the humor of which depends upon irrelevancy of the combined ideas; the Vice assigns reasons for his actions, which have logically nothing to do with the matter in hand.

Mischief will heal the broken heads of the minor Vices thus: „I xall help þe of þi peyne : I xall smytte of þi hede", etc., Man., 420.

Nought complains of pain:

„I haue such a peyne in my arme,
I may not change a man a ferthing". Man., 376.

Similarly, Ambidexter: „O, o, my heart, O my bum will break", K. C., 243. Likewise, Inclination:

„My little finger is spitefully sore;
You will not believe how my heel doth ache".

[T. T., p. 294.

Sedition refuses to tell his name:

„I am windless, good man, I have much pain to blow".
[K. J., p. 95.

Sensuality offers an excuse which has nothing to do with
the case: „I shall (come) anon had I wypt my nose", Nat., 1122ᴵ,
similarly, Riot:

„Fain of him would I have a sight,
But my lips hang in my light". Y., p. 13.

Haphazard: „For Conscience — — — Being hard heart-
ed was turned to a stone". A. V., p. 129. Other examples:
Hypocrisy: „Cover your head; For indeed you have need to
keep in your wit". L. J., p. 74.

Carnal Concupiscence: „That with talking and beholding,
their noses will bleed", M. M.², that is, simply the sight of a
white bosom will make a man's nose bleed.

Sin says to Damnation: „Your going grieves me so much
that the snot drops out of my nose". Money, Bii.

Sedition makes a senseless comparison:

„It is a great pity to see a woman weep
As it is to see a silly dodman (snail) creep,
Or, as ye would say, a silly goose go barfoot",
[K. J., p. 7.

Iniquity: „As is not betwixt this and hell", K. D., 238,
again: „I had rather then my new nothing, I were gon".
K. D., 1106.

Courage says inconsequently:

„The best are but shrews,
But I will not say so". Tide, Ciii.

On the other hand the Vice is often captious and logical
to a nicety, but only that he may pick a quarrel.

Nichol Newfangle calls Tom Tosspot a knave; Tom
replies: „Knaves are Christian men, else you were a Jew".
Thereupon Nichol remarks threateningly: „He calls me knave
by craft". L. W. L., p. 323. The same witticism occurs also
in F. El., p. 20: „He calleth me knave again by policy".

c) *Euphemism.* — *The Vice is very prone to the use of euphemism and circumlocution.* [Man., H., F. El., L. J., Y., Res., A. V., Confl., O., T. T., L. W. L., K. J., K. D., K. C., M., Tide.]

1. *For the sword.*

da pacem, Man., 699.
sheathe your whittle, H., p. 168.
sharp arguments, Confl., p. 50; similarly, p. 59.
wood-knife, L. W. L., p. 350.

2. *For the indecent.*

„my privyte" (sexual organ), Man., 414.
„doynge my nedynges", Man., 770.
„The kind heart of hers
Hath eased my purse", L. J., p. 79.
„Kiss where it doth not itch", K. J., p. 5.

3. *For hanging.*

„St. Andrys holy bende" (hangman's rope), Man., 614.
„A runnynge rynge-worme" (a skin disease, i. e. here the
    mark made by the rope), Man., 616.
„St. Patrykes wey" (Purgatory, put for death here by
    execution, Man., 600.
„Hanging stuff" (a criminal, a person to be hanged), Res.,
    I, 4, 34.
„leap at a daisy or put out the *i* of misericordia".
    Res., V, 2, 112, 113.
„To preach at Tyburn", Y., p. 15.
„Silk lace" (hangman's rope), A. V., p. 153.
„Hoist up the sail" (to hang), A. V., p. 147.
„the two legged mare" (the gallows), L. W. L., 352.
„for-letting my drink", A. V., p. 152.
„look through a rope", K. C., p. 216.
„to play sursum corda", Money, Di.
„hanging fare", Tide, Diiii.
„Beware thy arse break not thy necke", O., 679.

4. *Miscellaneous.*

„A relygyouse place" (brothel), Nat., 119[II].
„I beshrew thy father's son", F. El., p. 20.

„Thou shalt have a knaves skin" etc. (i. e., you are a veritable knave), F. El., p. 34; Similarly, T. T., p. 271.

„To heal his sore shins", Y., p. 17.

„Try you the trap", A. V., p. 132, (i. e., try it, make the venture.)

„Have with ye to Jerico", A. V., p. 138, (i. e., away with you).

„My melodie" (happy frame of mind)., Res., III, 6, 72.

„A Journey into Spain", L. W. L., p. 357 (to go to hell with the devil).

„They have sauce both sweet and sour", K. J., p. 10.

„Thou shalt have a mess of pease", K. D., 981.

„Like to make a burn", (i. e. at the stake), L. J., p. 74.

„I shall teach you your liripup to know" (manners), L. W. L., p. 322, Similarly, Tide, Biii.

„I cleft their cushions" (= heads), Tide.

„He hath increased a noble unto a ninepence", L. W. L., p. 344.

„Grumble seed" (money), Res., II, 2, 5.

„A bag of rie" (money), Res., V, 9, 70.

d) *Foreign words and phrases are used by the Vice in much the same way as by the other players. Expressions especially characteristic of the Vice are infrequent.* [Man., Nat., Res., T. T., M. M.².]

1. *Translations.* — Infidelity utters malicious side-remarks in Latin; as Mary praises her parents he says: „Puella pestis indulgentia parentum", M. M.², Bi; again, as she thanks him for his good advice: „Verba puellarum foliis leniora caducis", which he thus translates: „So fair a word truly changeth maids minds".

Sensuality translates „radix viciorum" ironically thus: „rote of all vertew", Nat., 841[1]: cf. Chaucer, *Nonne Preestes Tale*: Mulier est hominis confusio, Madame, the sentence of this Latin is, woman is man's joye and all his bliss".

Mercy, searching for Mankind, asks: „Ubi es"?

Newguise answers:

„Hic, hic, hic, hic, hic, hic, hic, hic,
þat ys to say, here, here, here", Man., 761, 762.

2. *Hybrid words: Mixture of Latin and English in phrases and sentences: feigning of foreign languages.* –

Mischief: „Corn seruit bredibus, chaffe horsibus, straw fyrybus", Man., 57, again:

„Here ys blottybus in blottis,
Blottorum blottibus istis", etc., — 667.

Nought: „in spadibus", 383, „hedybus", 384, „in nomine patribus choppe", 425, in manus tuas qweke", 502.

Nowadays: „He ys noli me tangere", 498.

Infidelity, singing:

„Salvator mundi Domine kyrieleyson,
Ita missa est, with pipe up alleluya,
Sed libera nos a malo and let us be at one", M. M.[2]

Avarice fittingly names Adulation „flaterabundus", Res., I, 3, 29, and „ait-aio", „negat-nego", 32.

Sedition mocks his companion who is reading the litany: „ora pro nobis", with „I beshrew vobis", K. J., p. 25.

Inclination attempts to escape from his enemies by pretending to be a foreigner: „Non point parle françois, non, par ma foy", and since this does not help his case, he tries again: „Ick en can ghene english spreken von waer", T. T., p. 277.

f) *Humorous Comparisons.* — [Man., Nat., Res.].

Occasionally the Vice shows a more genial side of his character by indulging in humorous comparisons. These good natured witticisms are rather infrequent.

Nought says depreciatingly of himself:

„I was never worth a pottfull of wortes sythyn I was
[borne,
My name ys Nought, I love to make mery", Man., 261—2.

Bodily Lusts makes perhaps the best joke of this type. When man commissions him to keep his dissolute company together, he says bravely:

„Mary, I shall do what I can thereto,
— — — But I shall tell you what,
I had leiver kepe as many flese,

Or wyld hares in an opyn lese,
As undertake that", Nat., — 642^{II}.

This recalls the speech of Puck, who also had to undertake a task impossible of accomplishment: „I would sooner keep fleas within a circle" etc., Ben Jonson's *The Devil is an Ass,* V, 2.

Avarice in *Respublica* is one of the best characterized of the Vices, his remarks possess this sort of humor in an eminent degree; the irony lies in the contrast between his name and his words:

„Nowe a wheale on such noses — — —
That so quicklye canne sent where hidden golde dothe [lye". Res., I, 3, 10.
„An ye looke at my bags, ye marre my melodie", III, 6, 72.
„I can goe nowhere now, in citie neither towne,
But Piers Pickpurse plaieth att organs under my gowne", [V, 2, 38,

and his moneybag he calls „a bag of rie", V, 9, 70, etc. etc.

g) *Under punishment the Vice is generally defiant, occasionally, however, he is witty.* — [K. J., A. V., Tide, M.]

Haphazard: „It would grieve a man having two plows going (i. e., it grieves one who has great enterprises in hand):

„Nay, stay and let the cat wink,
It is nought in dry summer for-letting my drink"

(i. e. what a pity to hinder my drinking right in summer time); Thereupon he makes his will:

„I will set, let, yield, permit and promise
All the revenues to you of my service", p. 152,

and protests once more:

„Why this is like to Tom Turner's dole:
Hang one man and save all the rest", A. V., p. 153.

Sedition regards himself a martyr who is already canonized: „Pray to me with candles, for I am a saint already", K. J., p. 99.

Courage pretends that he is not the guilty party whom Authority is seeking, and since he knows the criminal very well he offers to go and fetch him:

„So, sir, I thought you did me mistake,
I know right well the man whom you do mean,
... Yea, I will fetch him", Tide, Giii

Under similar circumstances Courage asks:
„Is there no man here that hath a cursed wife,
If he in my stead he shall end his life", Tide, Giii.

In *All for Money* Sin describes the stocks in which he had been sitting: „They be wormeaten, which shows them ancient to be", Money, Ciii.

h) *The Vice makes sport of his associates, especially those of inferior rank.* — [Res., O., K. D., L. W. L., Tide, M.].

Avarice 'says mockingly to Adulation: „What is your brainpan stufte with all, wool or sawdust?" Res., I, 4,2.

In O. the Vice is greatly amused because the fight Hodge is so concerned about his new hat: Ha, ha, he, mar his hat quoth he... for the blose he set not a pyn." O., 56.

Nichol Newfangle makes merry at the expense of the devil by playing on the motto-title of the play „Like Will to Like", he conducts the devil to Tom Collier, the clown. p. 314.

In *All for Money* Sin makes a great joke at the devil's expense; he says to the spectators: „You may laugh well enough that Sin and the devil be fallen out, But we will fall in again or ever it be long", Bii. Later as the minor Vices thank him for having comforted the devil, this important question occurs to him: „If the devil had died, who should have been his heir?" Biii.

In L. W. L. the drunken clown furnishes the Vice an opportunity for much merriment, p. 328, 329, 330.

Importunity calls Iniquity „Peter Pinchfist", 750, Iniquity retorts in like manner:
„What, John Coppersmith, otherwise called butterflie",
[K. D., 751.

Avarice has great enterprises under consideration, Adulation becomes importunate and officious, Avarice says to him:
„Who buzzeth in my ear so? What, ye saucy Jack.
What clawest thou mine elbow, prattling merchant?
[walk.

Ye flatterabundus you, you flyering claw-back, you,
You John-hold-my-staff, you what-is-the-clock, you,
You ait-aio, you negat-nego you." Res., I, 3, 32.

Courage inspirits the pickpocket: „If you are lucky, it is
well, if not — oh, then the hanging doesn't last long („It is
but an hours hanging"), Tide, Diiii.

IV. The Vice as a dramatic figure.

a) *Entrance.* — In the Moralities, the Vice enters, as a
rule, after the first principal act, that is, after the scene between man and the Good. His appearance on the scene is an
important event in the plot of the play. His entrance is equivalent to the introduction of new life and spirit, and is generally characterized by noise and bluster. As the various Vices
enter, however, in accordance with no fixed method, it will
be necessary to describe each entrance for itself.

Folly: „What ho! care away!
My name is Folly" etc.
Then he turns to the audience:
„Ah, sir, God give you good eve." W. C., p. 260.

Hickescorner: „Ale the helm" etc. He enters with a
sailor's expression and thereby motivates the following talk
about ships and travelling, H., p. 161.

Sensual Appetite: „Well hit" etc. and turns immediately
to Studious Desire: „Aha! now good even, fool" etc. F. El.,
p. 19.

Hypocrisy: „O, O, quoth he, keep again the sow." He is
summoned by the devil but as he sees the devil's shaggy
figure he pretends to be frightened. L. J., p. 63.

Riot: Hufa, huffa, who calleth after me? I am Riot, full
of jollity." He is summoned by Youth. Y., p. 13.

Iniquity: „Lo, lo, here I bring-a" etc. He enters singing
with Ismael and Dalilah, N. W., p. 168.

Avarice: „Now goddigod everychone" etc. He opens the
play and greets the audience, (goddigod = „god give you
good' [day]", Brandl) Solus, Res., I, 1.

Hypocrisy: „God speed you all" etc. He represents
himself as a priest. Solus, Confl., p. 45.

Ambidexter: „Stand away, stand away — — — — Harnessed I am" etc. He enters as a blustering soldier, miles gloriosus, and demands more room. K. C., p. 176.

Haphazard: „Very well, sir, very well, sir, it shall be done". He acts evidently as if there were some one behind the scenes, with whom he is talking, in all probability the devil, he says: „Who dips with the devil, he had need have a long spoon." Solus, A. V., p. 117.

Iniquity: „How now my maisters," etc. He opens the play and greets the audience. Solus, K. D., 35.

The Vice: „A, sirra! nay soft, what?" He is a sentry, overhears the conversation of the clowns. Solus, O., 1.

Inclination: „I can remember since Noe's ship" etc. Solus, T. T., p. 267.

Nichol Newfangle: „Ha, ha, ha, ha! now like unto like ... Stop gentle knave and take up your brother." He enters laughing, opens the play, plays a joke on a person in the audience by offering him a card, the knave. Solus, L. W. L., p. 309.

Infidelity:

„With high down, down and down a down a,
Salvator mundi Domine Kyrieleyson,
Ita Missa est, with pipe et alleluya,
Sed libera nos a malo and let us be at one."

He opens the play and sings mockingly a medley of church songs. Solus, M. M.[2]

Sin is represented as being vomited up by Pleasure as is directed by the stage-direction „Here he shall make as though he would vomit and Sin being the Vice shall be conveyed, finely from beneath as Pleasure was before," Money, Ciii.

Courage opens the play, sings(?) and describes his ship. Solus, Tide.

b) *Exit.* — The Vice makes his exit usually in the Moralities before the conversion of man, in this scene a Vice would manifestly be in the way. The dismissal of the Vice is not very skillfully or artistically brought about; as a rule, he is simply left out, as for example in W. C., H., L. J., etc. But sometimes the Vice is retained during the conversion-scene.

Since the conversion is the undoing of his work, he is, naturally, not pleased; he gives expression to his displeasure mostly by scolding and threatening, as in Y., an M. M.[2] In the Tragedies and later Moralities, on the other hand, a decided advance has been made in the dramatic art; the fate of the Vice, as it should, here forms an important part of the closing scene, justice is dealt him according to his deserts, as in K. J., A. V., T. T., etc. The various modes of the Vice's exit are classed thus:

1. *He simply leaves the stage.*

The Vice runs away of his own accord, Man., p. 68, H., p. 173, M. M.[2]

He marches off to London, W. C., p. 267.

He seeks a place of refuge, Nat., p. 146.

He goes away singing, F. El., p. 48, L. J., p. 89.

He is called out by Sir Lawrence to drink with him. Money, Ei.

2. *The Vice takes formal leave of the audience.* Y., p. 38, Confl., p. 115, K. C., p. 245, L. W. L., p. 357, W. W., p. 58, Money Ei (?).

3. *The Vice is led away to punishment (to prison, the gallows, etc.).* — K. J., p. 99, N. W., p. 176, Res., p. 357, A. V., p. 153, T. T., p. 297, Tide, G iii. In K. D., p. 403, fire is thrown on to Iniquity to indicate his destruction.

4. *The Vice is carried off by the devil in only one play.* — L. W. L., p. 357.

c) *Costume.* — A reliable source of information regarding the Vice's costume is the old wood-cuts; these are unfortunately rare, but possibly others may yet be brought to light. As a sort of frontispiece to *Hickescorner*, Dodsley's Old Plays, I, 147, and to *Jack Juggler*, Dodsley's Old Plays, II, 104 are two old wood-cuts, which represent the various persons in the respective plays. Hickescorner himself is dressed as a fine gentleman. Note that in this picture the Vice has no sword, although in the play he mentions his dagger, p. 171.

The picture of Jack Juggler is quite the reverse of that of Hickescorner; his dress consists of a short jacket with a belt, a loose cape about his shoulder and a slouch hat on his head. He has no sword. Jack Juggler is expressly designed in the list of players as the Vice. The play *Jack Juggler* is, it is true, a farce, not a Morality, but for this very reason, the picture in question is all the more valuable as showing that the costume of the fool was not the universal dress of the comical person, be he the Vice or the clown. Klein, Pollard and others apparently assume that, as a matter of course, all these characters appear dressed as fools.

In this connection mention may be made of the portrait of Tarlton, the famous clown, who, indeed, according to Fuller, was at one time a court fool of Queen Elizabeth's. He played the part of Derike in the *Famous Victories of Henry V*; in scene II special reference is made to his costume. According to the portrait, reproduced in Fairholt's *History of Costume*, and in *Tarlton's Jests*, Shakespeare Soc., 1844, the costume consists of a soft hat, a short jacket with a belt and long loose pantaloons; the pantaloons were regarded, in those days of buckle shoes and hose, as being especially countrified. A drum, a pipe and a large pocket or wallet at his belt completes Tarlton's out-fit. This is probably the „clown's suit" which he willed to his successor, Armin, see Jests, p. 23.

The stage-directions in so far as they contain descriptions of dress and equipment of the players, are a second source of information for the costume of the Vice, but the amount of information derived from this source is limited: Heywood's *Play of Love* (Brandl, *Quellen*, p. 200): „Here the Vice cometh in running suddenly about the place among the audience with a hye copyn (huge coppyr, Fairholt) tank on his head full of squibs fired" etc. Belial in the *Castle of Perseverance* was similarly accoutred, see above p. 40. The *King Cambyses*, Dodsley, IV, 176, gives a detailed description of Ambidexter's out-fit: „Enter the Vice with an old capcase on his head, an old pail about his hips for harness, a scummer and a potlid by his side, and a rake on his shoulder." He enters in full armor, his costume in keeping with the warlike tone of the play, but his armor is grotesque; his helmet is an old hat-box, his coat of mail an old pail, his sword a ladle, his shield a potlid, his spear a rake.

The situation in O. is similar to that in K. C., that is, a war is imminent, but whether Ambidexter enters armed or not, is not indicated. He mentions his sword, 60, and says in his song „to wares I muste", 672.

In *All for Money* nothing is said about the out-fit of the Vice excepting his sword. This is the more surprising as the costumes of the other figures are described with minuteness, for example,

> Money comes in white and yellow,
> Science, as a philosopher
> Theology, as a prophet,
> All for Money, as a magistrate.

The two minor Vices are dressed as devils: „Here cometh Gluttony and Pride dressed in devils apparel", M., B ii and „Damnation shall have a terrible visard and his garments shall be painted with flames of fire", M., B i; similarly Judas and Dives.

That the Vice in this play asks the devil for a piece of his tail and his mask, indicates nothing regarding the Vice's costume, at any rate he does not secure the courted articles.

A change of Costume is sometimes provided for according to the stage-directions. Thus in O. the Vice appears first as a messenger from God and as a herald of war, then dressed as beggar, and, finally, as Revenge, but it is only once indicated how he dresses himself for the different rôles; the direction for this last entrance is: „Vice entereth with a staff, a bottle or dish and wallet", followed by the further direction: „Put off yᵉ beggars coat" etc., p. 532.

In W. W. Idleness appears once as a rat-catcher, p. 39, and once as a priest, p. 57, each time in order to deceive the constable.

Avarice in Res., carries on his back, but on the inside of his coat, a number of pockets or bags, I, 1, 46. Later he is forced by Verity to turn this coat inside out and expose his secret pockets, V, 9, 91.

In the case with Infidelity in M. M.[2] the change of costume is very carefully provided for; as he first goes to Mary Magdalene he puts on a special dress: „Infidelity put on a gown and a cap", Ci, and asks Pride if it sits well, Cupidity is of the opinion that there is yet one thing necessary, Infidelity should do something to disguise the foolish expression of his face. Mary as well as others remark his changed appearance. On another occasion he says: „Among the pharisees I have a pharisees gown, And among publicans and sinners another I use", Ei, and accordingly, as he prepares to go to the banquet at Simon's house, he puts on the dress of a pharisee: „Well remembered, yet I must provide a garment, Against that I come to master Simon, About the which the precepts of the Testament must be written", Eii.

Allusions to dress. — Another source of information for the nature of the costume, is furnished by the references to it in the text of the play; these, however, in order to be reliable, must in each case be clear and unmistakable. If, for example a person says „I have played the fool", one is not necessarily justified in concluding that he is dressed as such.[1]) In the words of the players themselves, however, are found occasionally references especially to particular articles of costume.

1. *Spectacles.* — In K. J., p. 30. Dissimulation says: „With my spectacles vadam et placebo"; in T. T., p. 269, Sturdiness wears a pair of large spectacles for the purpose of frightening the Vice. Spectacles are a peculiarity of the costume of Nobody: cf Shakespeare Jahrb. XXIX, XXX).

2. *The Sword.* — The sword or the dagger is mentioned more frequently than any other article, especially where the

---

[1]) The dramatists make use of many expressions which are to be understood as figurative or as mere epithets, for example: „fancy in a fool's case", Mag., 1058 (case = skin: Cf. „In case my lady do threaten my case", A. V., p. 123). „Play the fool without a visor", Mag., 1192. „Who spake to þe, foll?" Man. 132. „I pleyde so longe þe foll", Man., 264 (Otherwise, Brandl, *Quellen*, XXXIII). „I gave him a blow with a foxtail." K. D., 192 (i. e. to get the better of one to make a fool of him, Halliwell). „Learn to keep that cockscomb of thine", O., 154. „For a wyntur corne threscher, ser, I have hyryde", Man., 54. (Brandl concludes from this that Mischief was dressed as a farm hand, *Quellen*, XXXIII, but this does not necessarily follow. The passage in question is merely a play on words, see above, p. 107, γ.

Vice indulges in threats or wants to fight. Sin is born armed with a sword, Money, Bi; See below.

That the sword of the Vice was of wood, can be established by at least one citation.

Nichol Newfangle says:

„Lest I stick you with this wood-knife,

— — — — — — — — —

Body of me, they have ta'en away my dagger",

[L. W. L., p. 350.

The Hangman's rope. — Newguise enters, as a gallows-bird or an escaped captive:

„... þe halter brast a sondre; Ecce signum,

The halff ys abowte my neke", Man., — 603.

Mischief likewise, as is evident from Nought's remarks:

„Me semyth ye have scoryde a peyr of fetters."

[Man., 628.

Under similar circumstances Riot enters, Y., p. 15. He explains that he has just come from the gallows; whether he wears a rope or fetter is not mentioned.

3. *The Bridle.* — In T. T., p. 278, a bridle, the recognized symbol of self-restraint, is put on Inclination, the Vice, thus showing how tendencies to evil must be curbed.

4. *Masks.* — In none of the extant plays is the Vice ever spoken of as wearing a mask. A mask is mentioned only once, in W. W., p. 38, where Irksomeness is deprived of his head, that is, his mask. Irksomeness, however, is hardly to be considered a Vice.

5. *The fool's costume.* — Indubitable references to the fool's costume, in connection with the Vice, are not frequent. Neither the Moralities nor the Tragedies justify the opinion that the fool's costume was so generally used as many, Pollard, Klein and others, suppose. It is only in the two earlier Wit plays that the fool's costume comes into prominence, but it is not there the dress of the Vices but of their fools. Idleness and her fool dress the sleeping Wit in a fool's costume.

6. *Fashions.* — Haphazard says of himself:

„Yet a proper gentleman I am . . . .

Yea, that ye may see by my long side-gown."

[A. V., 118.

He speaks ironically, however, for he represents himself as a student, teacher, fisher, butcher, thief, hangman, etc.

In Y. Pride would be a gentleman in spite of his shabby clothes, p. 25. „Light apparel" means generally, simple citizen's dress, as opposed to uniform. In K. J., p. 34 it refers to the dress of a rogue; and in *Albion Knight* to that of a fool, as the use of the word „foxtail" indicates. In M. M.[1] Curiosity is a gallant cavalier, a dandy, a „Kleidernarr", according to Brandl, *Quellen*, XLI. In M. M.[2] Infidelity enters first as a cavalier, then as a pharisee.

7. *Deformities*. — There are very few references which indicate that the Vice is represented as physically deformed. In O. the clowns call the Vice a hedgehog, „this little hourchet", 46, but this is probably only an epithet. In K. C., the Vice is perhaps hunchbacked like Punch, at all events his appearance makes a strong impression upon the clowns. Huff says of him: „Such a deformed slave did I never see," p. 178. The appearance of Infidelity must have been very revolting; he was squint-eyed, as is evident from the speeches of the other players. Cupidity speaks of his „foolish countenance" and says: „Thou lookest like one that had lost his remembrance," and carnal Concupiscence says expressly: „With one eye over much thou useth to wink." M. M.[2], C i.

Thus it is evident, that the Vice enjoys the greatest freedom in the matter of dress; he is not confined to any stereotyped costume; he represents the soldier, beggar, priest, vagabond and fool; and in fact, seems to prefer the irregular and grotesque. His costumes show moreover, that the Vice is a person of great versatility, but the opinion that he is always or usually dressed in a fool's costume, has absolutely no justification. Such a supposition rests upon an entirely false conception of the Vice-figure.

d) *Song and Dance*. — In all the extant Moralities and Tragedies, excepting Nat., H., Money, the Vice either sings, or mention is made of songs. His singing usually is simply an expression of mirthfulness; occasionally, however, it is more closely connected with the subject of the plot, or assumes some particular form.

In Man. Nowadays proposes „a crystemes songe“, and Nought invites the spectators to join with them in the singing:

> „Now I prey all þe yemandry þat ys here
> To synge with us with a mery chere,
> Yt ys wretyne with a coll“, Man., — 324.

This song is a shocking parody of an anthem.

This motif is especially developed in the F. El.; soon after his entrance Sensual Appetite trills the following words, which clearly show him to be a gay and genial person:

> „With a huffa gallant sing terl on a berry,
> And let the wide world wind,
> Sing, frisky jolly, with hey trolly lolly,“ p. 20.

He next proposes amusements, namely dancing, laughing, and merry songs. Later he goes to outfind singers and dancers. Since an instrument is lacking, he sings without accompaniment. For his performance on this occasion he is highly praised by Ignorance. The stage-directions for this scene are more definite than usual: „Then he singeth and danceth withal and evermore maketh countenance according to the matter; and the others answer likewise“, p. 47. A similar direction is found in L. W. L, p. 315.

In A. V., p. 122, 134, Haphazard sings and dances with the players of lower rank, servants etc. Nichol Newfangle does the same especially with the clowns and the devil. He requires also a guitar, L. W. L., p. 315. In A. V. the songs play upon the name of the Vice Haphazard, in L. W. L. upon the title of the drama. In N. W., p. 168 the Vice sings with Ismael and Dalilah.

In L. W. L., p. 332, Nichol Newfangle in his song refers to preceding events in the play: „Now these knaves are gone“ etc. Similarly, „the Vice“ in O., 648, 850, Confl., p. 78, Tide, Diii, Eiiii. In a song Nichol praises the things he has furnished to Tom and Ralf, namely, the hangman's rope and the beggar's staff and wallet etc. He sings „Trim merchandise“ etc., etc. L. W. L. p. 344. Avarice and the other Vices in Res. sing appropriately: „Hey, noney, nony, houghe for money“ etc. Res., p. 318.

e) *Fighting.* — Quarrelsomeness is one of the most prominent traits of the Vice. This motif is lacking in F. El., L. J., Y., K. J., Res., M. M.², Money.

1. *The threatening and the abuse of the Good, at the hands of the Vice, has already been discussed, see above, p. 83, d.*

2. *The Vice is only too friendly disposed towards man to wrangle with him, and when he does quarrel with him, it is generally only a temporary matter.* [Man., W. C., N. W., T. T., Tide.

In Man. the three minor Vices receive a vigorous beating with a spade by Mankind, as a punishment for their rough teasing; strangely enough they offer no resistence, p. 52.

In W. C. occurs a good fighting-scene. The Vice, who professes great ability as a fighter, tauntingly challenges man to a bout with the sabers. The fight turns out favourably for Manhood, as is to be expected, for Folly's only purpose in proposing the fight is to get acquainted with him, p. 261—2.

Hickescorner earnestly, but greatly to his own disadvantage, seeks to allay the strife between Imagination and Freewill. As in the wellknown story of the Irishman and his wife, the opposing parties lay aside their own quarrel, in order to make a common attack on the peacemaker, Pity; Hickescorner gets a broken head, p. 168.

In N. W., after Iniquity, with Dalilah's aid, has won all of Ismael's money, he quarrels with the girl about their common winnings. He accuses her of cheating and finally gives her a box on the ear, p. 172.

In T. T., fisticuffs take place between Lust and Sturdiness and Just in the scene before the entrance of the Vice. While Lust and Sturdiness are boasting about their treatment of Just, the Vice enters and tries to outdo the two men in bragging; but as soon as they threaten him with a sword, and look fiercely through their spectacles, he shows the white feather, p. 269.

Courage strikes the courtier soundly on the back, to attract his attention, says, by way of excuse, that he has made a mistake in the person, Tide, E i.

3. *With the clowns.* — [K. C., A. V., O., L. W. L.] — The appearance of clowns in the later plays occasions the Vice much surprise and trouble. Vice and clowns can in no way live in peace with one another, their quarrels generally turn out badly for the Vice.

α) *The clowns attack the Vice.* The three ruffians, Huff, Snuff and Ruff, are engaged in a lively conversation about the war, Ambidexter mingles in the talk, but they soon become suspicious of him. Huff says to Ruff: „Do you know him?" Ruff answers: „No! I never see him before", Snuff proposes that they shove him against the wall; Ambidexter, of course, gets angry: „Ah, ye knaves, I will teach you how to deride me." The stage-direction here requires that he shall give them a good trouncing: „Here let him swinge them about", K. C., p. 179. Although Ruff and Snuff beg for mercy, he begins pummeling them again; but this time they draw their swords. Ambidexter now declares himself ready for peace: „Let us agree", and they shake hands.

Ambidexter has still another fight; Hob and Lob, two rustic clowns are on the way to market. Ambidexter meets them as they are discussing the cruel acts of the king. He leads them on to make some treasonable remarks, and then threatens to report them. In order to prevent his doing this, they give him cakes and a goose as bribes. Hob and Lob then begin to quarrel with each other. Ambidexter observes his opportunity and says aside: „I will cause them to make a fray", and then to Lob: „Yea, Lob, thou sayest true, all came through him." Then Hob and Lob come to blows, as the Vice desired. The stage-directions are: „Here let them fight . . . the Vice set them on as hard as he can; one of their wives come out, and all to beat the Vice, he run away." Hob and Lob make peace, but the fisticuffs are not yet over. Marian, the wife of one of the clowns and Ambidexter engage in a rough and tumble fight, and this, indeed, for the sole purpose of creating merriment: „Here let her swinge him down and he her down, thus one on the top of another make passtime." K. C., p. 224.

The circumstances in O. are the same as those we have just considered in K. C. The Vice overhears Rusticus and

Hodge discussing the king's affairs; he finally joins them, but
they are greatly incensed at his presence; they are of the
opinion that, as he is such a little fellow, they could soon
make away with him. In the scuffle which follows, the Vice
staves in Hodge's new hat: „Hold, good master, ye mar my
new hat". This strikes the Vice as being so funny that he
is compelled to laugh, his merriment continues until Hodge
inadvertently distorts the Vice's pretended name, changing
„Patience" to „Past shame". This makes the Vice so
angry that he resolves within himself to take revenge, which
he does by causing a fight between the two clowns; to one of
them he says: „Sirra, you, goodman Rusticus, marke what
I saye — — — this dyd I see, a hoge of thyne wearyed to
be", 108.

Rusticus is naturally aroused and wants to know whose
dog it was that worried his hog; the Vice says slyly:

„Ha, ha, ha — — — it was a very shame,
For thy neighbor to let it".

The clowns then begin to quarrel; Hodge pretends, by
way of an excuse, that his neighbour's hog had ruined his
garden: „My rye and my otes, my beanes and my pease".
They soon come to blows; „— — — but Hodge smit first;
and let ye Vice thwacke them both and run out", p. 498. This
passage recalls a similar one in *Nature*, where the Vice, during
the fight between Man and Reason knocks the latter on
the head.

In A. V. Mansipulus, a servant, with much foul language,
attacks the Vice, Haphazard, the latter ably defends himself;
but finally Mansipula takes a hand in the difficulty; this is
too much for the Vice: „Nay, sure I have done when women
do speak", p. 121.

β. *The Vice attacks the clowns.* — The two clowns, Tom
Tosspot and Ralph Roister, contend with each other, as
to which is the greater rascal; Nichol Newfangle acts as
umpire: „I will sit in this chair and give sentence on the
same", etc. Nichol begins by blaming them both, first, because they have the audacity to remain before his judgment
stool, with their hats on, secondly because they address him

as „Nichol" simply, instead of „Master Nichol Newfangle". He proceeds energetically to teach them better manners: „I shall teach you both your liripup to know". Thereupon follows this stage-direction: „He fighteth", and again: „He fighteth again", L. W. L., p. 322. Later Nichol Newfangle tries to play a practical joke on these same clowns, and succeeds so well that they turn upon him, they give him a sound thrashing and take his wooden dagger away from him, p. 350.

4. *With his accomplices.* — Quarrels between the Vice and the minor Vices are infrequent, but sometimes the Vice adopts stringent measures in order to establish or to maintain his own priority. Generally these affairs result in nothing but words, as for example between Envy and Pride, Nat., 824[II], seq., and in Money, Bii, iii, especially in the latter case, Sin lays claim to precedence among the Vices. In K. D. the Vice plans an attack, and that without any apparent cause, on Partiality and Importunity: „With one blow on this side and one on that" etc., 179. But they are not willing to fight; again he tries to quarrel with them, because they take exception to his abusive language.

In Confl., Hypocrisy tries to play the rôle of the Good, with the purpose probably of testing his accomplices, but they see through his pretentions, and give him a sound thrashing, p. 52, seq.

In Tide, Courage attempts forcibly to retain a minor Vice, who is about to go away, Courage draws his sword, a third Vice intercedes and peace is soon restored. Much more serious is the contest between Courage and Help, he demands a portion of their ill-gotten gains, Help refuses to give it him, whereupon Courage seizes him. The stage-direction is as follows: „And fighteth to prolong the time while Wantonness maketh ready", Eii. Thus fighting is used to fill out the time.

f) *Cowardice.* — [Man., L. J., K. C., O., L. W. L., T. T., M. M.[2], Tide].

Fully in keeping with the traits of the Vice's character as already described, particularly, quarrelsomeness and boastfulness, is that of cowardice.

1. *The sudden appearance of the Good* causes Mischief and the other Vices to take to their heels, Man., p. 68.

Infidelity is at first greatly frightened when Christ enters. He says:

„Benedicite, Art thou come with a vengeance?“ M. M.[2]
[Fii.

In Tide, Courage and Greediness pretend not to see Christianity, Fiii, and later when Christianity appeals to God, Courage simply goes away, Fiiii. As a rule the Vice is afraid of the Good only where, at the end of the play, he is about to receive punishment; for example, Inclination tries to escape from Sapience, naively remarking that he has no heart to meet him: „My courage is spent, I have no more“. T. T., p. 277. In Tide, Giii, short work is made of the braggart, Courage; the stage-direction is simply as follows: „Courage catches him“. How effectively this was done, may be judged from Courage's cry of consternation: „O, God's passion, wilt thou break my neck?“

2. *In the presence of the devil.* — Hypocrisy in L. J., p. 63, and Nichol Newfangle in L. W. L., p. 311, show temporarily signs of fear because of the devil's frightful appearance, but as soon as they see that the devil is not a bear or a hog, they confidently approach him.

3. *With the Clowns.* — The Vice in his conflicts with the clowns proceeds with extreme caution. In O. the Vice wishes to punish Hodge and Rusticus, but it occurs to him, that, in this case, it is a matter of two against one: „two is too many“, 78. He considers the matter and finally comes to the conclusion that it is „good sleepinge in a hole skynne“, 103.

Often in these fighting-scenes the Vice coward-like runs away. Ambidexter, for example, describes afterwards the conflict between Meretrix and the ruffians:

„I may tell you I was in such a fright . . .
I made no more ado, but avoided the thrust,
And to my legs began for to trust“. K. C., p. 186.

On two other occasions he does the same:
„The Vice run here for fear“, p. 185, also p. 222.

4. *By means of the spectacles and a sword* Inclination is quickly silenced, T. T., p. 269. Ambidexter imposes upon the clowns in a very domineering manner, but as soon as they draw their swords he cowers: „O the passion of God, I have done", K. C., p. 180.

g) *The Vice as a braggart.* — [F. El., K. J., K. C., K. D., O., T. T., L. W. L., M. M.², M., Tide].

The exploits of the Vice are, according to his own words, for the most part grotesque and absurd.

Thus Sensual Appetite describes his action in the fight:
„Yea I have slain them every man,
Save them that ran away",

but it soon becomes apparent that all had run away except one, whose leg he cut off; he would have cut off his head, but some one else had already done that, F. El., p. 41.

Similarly; Courage:
„I was twice smitten to the ground
. . . But I had no wound". Tide, Bi.

Ambidexter:
„I am appointed to fight against a snail,
To be a man my deeds shall declare". K. C., p. 176.

Iniquity brags in the presence of his companions, how he succeeded in putting Equity to flight:
„But the knave was glad to take hys flyght,
He durst tary no longer in my syght,
By this you may know I was a bolde man", K. D., 528.

The fact of the matter is, that as we know from the play, it was all just the reverse; Equity's departure was wholly voluntary, he left because of disgust, at the same time warning Iniquity what he might expect.

Sedition brags of his strength:
Ye cannot subdue me
„though were as strong as Hector and Diomedes",
[K. J., p. 9;

Inclination tries in the same way to impose upon the clowns:
„It was I . . . Which brought to confusion both
Hector and Alexander" etc., T. T., p. 268—9.

In L. W. L., p. 321, Nichol Newfangle assumes the dignity of a judge: „Where learned you to stand capped before a judge?" Ambidexter assumes the dignity of a great man, K. C., p. 233, and Sin does the same, *All for Money*, Bii. In O., the Vice pretends to be a great warrior:

„But in this stower who beare the fame
But onley I?" O., 664.

h. *Sensation.* — That the Vice possessed a spirited manner of playing, and that he indulged freely in gestures, and made faces, and played all sorts of pranks, is rather suggested than directly stated in the plays. The stage-directions and references in the speeches of the players illustrating this point are not abundant. [Man., F. El., N. W., K. J., K. C., Confl., T. T., L. W. L., M. M.², Tide].

1. *Singing accompanied by gestures*, F. El., p. 47,

2. *Court-scenes*: Court of Mischief, Man., 651 ff., L. W. L., p. 321, seq.

3. *Playing dice*: „He casteth dice on the board", N. W., p. 169 (In Y., p. 34, various games of hazard are mentioned).

4. *Mimicry*: Greedy Gut opens his mouth wide, the Vice does the same: „Gape and the Vice gape", T. T., p. 273.

The Vice plays that he is a horse, whinnies, and kicks when bridled, T. T., p. 278, 281, 297, 299. The Vice plays the priest, Confl., p. 45, 51, K. J., p. 25, 45, 46.

5. *Noise*: „Sedition extra locum" creates a disturbance: „Alarum tro, ro, ro", etc., „thomp, thomp" etc., K. J., p. 53, similarly, Infidelity and the expelled devils, M. M.².

6. *Stupid tricks*: Ambidexter while trying to help deck the table, stumbles and falls with a dish of nuts: „Let the Vice set a dish of nuts and let them fall in bringing them in". K. C., p. 234.

Courage enters behind a courtier and slaps him on the back, to attract his attention: „And smiteth the gentleman", Tide, Ei: Cf. „Well hit", F. El., p. 19.

7. *He is carried on the back of others*, K. J., p. 31.

i) *The giving of information is one of the most common traits of the Vice, there is scarcely a play in which it is wholly lacking.*

1. *The Vice gives his name, either voluntary or on demand.* (*Lacking in H.,* N. W.).

„My name is Folly", W. C., p. 260.

„I am Riot", Y., p. 13.

„My name — — — I have forgot it

Ha, ha, now I have it,

My name is Ambidexter", K. C., p. 177.

„My very true unchristian name is Avarice",

[Res., I, 1, 11,

„Courage contrarious or Courage contagious,

That is my name", Tide, Ai.

Other examples: F. El., p. 21, Man., 112, K. J., p. 8, K. D., 358, A. V., p. 118, M. M.[2] Bii.

In O., the Vice has two distinct names: „Amonge the godes celestiall, I Courage called am", 207, „and I, Revenge", etc., 1047.

Often the Vice gives an assumed name for the purpose of deception.

Instead of „Folly and Shame" he pretends to be merely „Proper Folly", W. C., p. 264.

The devil names Hypocrisy Friendship, and this false name is used by the Vice himself in the presence of Youth, Y., p. 68.

Riot calls himself Friendship, L. J., p. 71.

Pride calls himself Worship, Nat., 939[I].

Avarice: „I will my name disguise

And call my name Policie instede of Covetise".

Res., I, 1, 22.

Infidelity tells Mary that his name is Prudence, M. M.[2]

In similar manner the names of the minor Vices are distorted. It is generally the principal Vice himself, who attends to this.

2. *The Vice often gives information concerning his lineage, character, vocation, etc.* — [Man., F. El., W. C., L. J., K. C., T. T., Confl., M. M.[2], Tide].

From a mere reference to a vocation, however, the conclusion that the Vice wore a costume corresponding to the same is not warranted. The Vice may pretend to be a farmer, butcher etc., but generally this has no further purpose than to give point to a witticism or a satirical allusion.

Mischief alludes to the words of Mercy:
„For a wyntur corne threscher I am hyryde“, Man., 54.

Folly:
„Yea, sir, I can bind a sieve and tink a pan“ etc., W. C., p. 261. He is therefore a tinker; he is also „a servant of the law“, p. 262.

Hypocrisy:
„For by my occupation I am a butcher“. L. J., p. 63.
He at first mistook the devil for a hog.

Ambidexter explains the meaning of his name:
„I signify one
That with both hands finely can play“. K. C., p. 177,

Likewise Courage:
„Courage contagious,
When I am outrageous
In working of ill, and Courage contrary,
When that I do vary,
To compass my will“. Tide, Ai.

Hypocrisy describes his nature:
„We Mercurialists, I mean hypocrites“, Confl., p. 47.

Similarly Sensual Appetite, F. El., p. 21, and Infidelity, M. M.[2], Bii.

Inclination, recalling his past, says:
„I can remember since Noe's ship“ etc. T. T., p. 267.

Similarly, Nichol Newfangle, who had been before his birth a pupil of Lucifer, L. W. L., p. 309.

Nought says of himself depreciatively:
„For I was never worth a potfull of wortes“, Man., 261.

Hypocrisy boasts that he is a good servant of the devil:
„Trudge, Hypocrisy, trudge
Thou art a good drudge,
To serve the devil“. L. J., p. 69.

136

3. *The Vice narrates his experiences as a traveller.* [W. C., H., L. W. L., M.].

Folly has travelled over all England and has visited the brothels of London as well as the English cloisters, W. C., p. 263.

Hickescorner recounts a long list of countries which he has visited, H., p. 1612.

Nichol Newfangle has travelled over the whole world, L. W. L., p. 310.

Sin, between his first and his second entrance, has visited many nations:

„I have been since I was here in many a nation“.
[Money, Ciii. Cf. also K. J., p. 8, 9.

4. *Of experiences before his birth.* — [T. T., L. W. L., M.].
Inclination:

„I can remember since Noe's ship . . .
Since Paradise gates were watched by night“ etc.,
[T. T., p. 267.

Nichol Newfangle:

„First, before I was born, I remember very well,
That my grandsire and I made a journey into hell“ etc.,
[L. W. L., p. 310.

Sin tells about his birth:

„I was afraid of nothing but only my dagger,
Lest at the time of my birth it would have sticked my
[father“, Money, Bi.

5. *Of his criminal experiences.* — [Man., H., Y., M. M.², Money].

Hickescorner travelled with a great company of all sorts of criminals, H., p. 164.

Riot:

„I came lately from Newgate“ — — —
„Verily, sir, the rope brake,
And so I fell to the ground, . . .
By the way I met a courtiers lad,
And twenty nobles of gold in his purse he had“. Y., p. 15.

Mischief likewise:

„Of murder and manslawter I have my bely fyll“ Man., 626.

Infidelity:

„Much woe had some of us to escape the pillory".

[M. M.², Aiii.

Sin also similarly, Money, Ciii.

6. *The Vice announces his plans and that generally without attempting to be comical or satirical.* This motif is not especially prominent in Man., Nat., H., F. El., L. W. L., T. T., N. W.

Folly, aside:

„I shall draw him such a draught of drink,
That Conscience he shall cast away". W. C., p. 265.

Hypocrisy to the devil:

„I warrant you, let me alone,
I will be with Iuventus anon

— — — — — — — — —

I will infest him with wicked company", L. J., p. 68.

Infidelity would deal similarly with Mary Magdalene;
M. M.², Bii.

Ambidexter: „by the mass, I will cause them to make a fray", K. C., p. 221. Likewise the Vice in O., 105:

„Well forwarde I will for to prepare
Some weapons and armour" etc., O., 6.

Avarice:

„I have a hive of humble bees swarmynge in my brain",

[8, — — —

„And nowe ys the tyme come that — — —
Een to make up my mouth and to feather my neste".

[Res., I, 1, 29, 30.

Ambidexter:

„For while I mean with a soldier to me,
Then give I a leap to Sisamnes the Judge" etc.,

[K. C., 177.

Iniquity:

„I must myself bestir,
In my wrath and ire,
That they shall come no more
Which have me sore vexed". K. D., — 561.

Hypocrisy gives a detailed account of his activities,

Courage announces the destination of his ship:
„Therefore we sail
To the devil of hell". Tide.

j) *The Vice addresses the audience.* —

The barrier between the audience and the players did
not then exist to the same extent as now, least of all for
the Vice.

1. *The Vice to the audience in general.*
„Lo, sirs". W. C., p. 266.
„Ah, ah, sirs". W. C., p. 265.
„Well, sirs". Tide, Diiii.
„Ha, ha, lo, masters", K. D., 524.
„Masters", Money, Ei.
„My masters", L. W. L., p. 357.
„My masters". K. C., p. 245.
„Make room, sirs". F. El., p. 20.
„Take heed that none of you hit my left heel". Money,
[Ciii (i. e. as he kneels).
„You may laugh" etc., Money, Bii.
„Syrs, who is there that hath a stoole?
I will buy it" etc., K. D., 109.
„Is there no man here that hath a cursed wife". Tide Giii.
„Is theyr neare a man that a servant doth lacke". O., 1053.
Nought invites the audience to join in the singing:
„Now I pray all the yeomanry, that is here,
To synge with us with a mery chere". Man., — 323.

Ambidexter makes a pun on his name and at the same
time calls the attention of the audience to the action of
the play:
„How like you Sisamness for using of me?
He played with both hands" etc., K. C., p. 209.

Similarly, Sin calls attention to the influence of money
in the world: „Do you not see how all is for money?" Money,
Ei: Cf. also Courage, who makes a pun on his name and
on the word „incourage".

2. *To the women.* —
Ambidexter;

„How say you, maid? to marry me will you be glad?"
[K. C., p. 232.

Inclination:

„Nay, for the passion of me, be not so moved". T. T., p. 287.

Nichol Newfangle:

„How say you, woman, you that stand in the angle
Were you never acquainted with Nichol Newfangle?"
[L. W. L., p. 309; again:

„What sayest thou to it, Jone with the long snout?"
[p. 317: again:

„How say you little Meg", p. 355.

Courage:

„How say you wives", Tide, Ciii; again:

„How say you virgins", Tide, Dii.

The Vice:

„God morrowe, mystres Nan! — — —
— — — Nay, may I be so bold at your lyppes to have
[a lycke" etc., O., 871—873;

again: „gentle woman", 1054.

again: „Ye, faull to it good wyues", 1089.

3. *To individuals.* —

Upon entering, the devil addresses Nichol Newfangle,
who, as if he were not the one addressed, says to a spectator:
„He speaketh to you, sir" etc., and at the same time points
out a person in the audience, L. W. L., p. 311.

Similarly asks Hypocrisy, when threatened by Tyranny,
Confl., p. 52.

Nichol Newfangle on entering offers a playing card to
one of the spectators, at the same time making a comical
allusion to the proverb „Like Will to Like", he says: „Stop
gentle knave and take up thy brother" (the card is called
knave) L. W. L., p. 309; again: „Why, gentle boy, how likest
thou this play?" p. 355.

4. *To the pickpockets.* —

Ambidexter:

„But is not my cousin
Cutpurse with you in the meantime?
To it, to it, cousin; and do thy office fine". K. C., p. 209;

again: „But cousin, — — —

Frequent your exercises, a horn on thumb" etc., p. 235.
Other examples, A. V., p. 124, 129; O., 676, 1120, L. W. L., p. 334.

k) *Reflecting the action, somewhat in the manner of the
ancient chorus, is a marked trait of the Vice (lacking in H.).*

1. *The Vice tells what is going to happen.* — [Confl., Res.,
K. C., A. V., O.].

Hypocrisy:

„What shall become of this foolish goose, I mean Philo-
[logus — — —
He shall not long continue so" etc., Confl., p. 115.

Ambidexter:

„He will not be quiet till his brother he killed" etc.,
[K. C., p. 215; again:
„If the king use this gear he cannot live long", p. 218;
and „I lay twenty thousand pound
That the king die by some wound", p. 244;
The wounded king enters and dies.

Haphazard describes a number of impossible conditions
under which Appius may obtain Virginia, A. V., p. 130.

Avarice says sententiously:

„A daughter eke he (time) hath, called Verity — — —
She bringeth all to light, some she bringeth to shame".
[Res., III, 6, 85. (Verity enters for the first time, V, 3).

The Vice, aside, as soon as Horestes has obtained the
king's consent:

„In revenging the wronge his mynd he hath set,
It is not Idumaeus that hath poure to let
Horestes from sekinge his mother to kyll". O., 256.

2. *The Vice likes to talk about his own deeds.* — [Man.,
Nat., K. C., A. V., L. W. L., Tide].

Mischief:

„Alasse þat euer I was wrought — — —
I, Mischief, was here at þe begynnynge of þe game,
Ande arguyde with Mercy", Man., 402.

Sensuality:

„I have brought thys man to his old gyse", Nat., 322[II].

Haphazard describes how much he had to do for judge Appius:

„Why — run sir knave, call me Claudius,
Then — run, with a vengeance, watch Virginius,
Then — ride, sirrah; is Virginia at church" etc., A. V.,
p. 150.

Ambidexter comments upon his capability as a deceiver:

„Marry, sir, I told him a notable lie.
Thereby you may perceive I use to play with each
[hand", K. C., p. 215.

Courage, likewise, Tide, Dii:

„My words have set her in such a heat".

Nichol Newfangle describes his fear of the devil, L. W. L., p. 317; likewise Ambidexter relates how frightened he was, while the clowns were fighting, K. C., p. 186.

3. *He exhibits much satisfaction at the results of the action of the play and particularly the success of his own malicious mischief-making.* — [K. J., K. C., A. V., Confl., T. T., M. M²., Tide].

Hypocrisy: „Such chopping cheer as we have made, the like hath not been seen", Confl., p. 115.

Sedition: „Is not this a sport? K. J., p. 65.

similarly, Inclination, T. T., p. 276.

Haphazard: „By our Lady Barefoot, this bakes trimly". A. V., p. 137.

Ambidexter: „Doth not this gear cotton". K. C., p. 215.

Hypocrisy: „Ha, ha, ha! marry, now the game begins". Confl., p. 65; similarly also

Infidelity: „Ha, ha, ha! laugh, quod a". M. M².

Courage:

„Now you may see how Courage can work,
And how he can incourage both to good and to bad",
i. e. the Courtier and Greediness:

„Ah, sirrah, I cannot choose but rejoice", Tide, Ciii.

4. *The Vice reports the entrances and exits of the various players.* — *Lacking in* W. C., H. A few examples will suffice.

Mischief on the sudden appearance of Charity:

„Tydyngs, tydyngs, I haue a spyede one;
Hens with yowur stuff", Man., 708.

Sensuality calls man's attention to the arrival of Pride (Pry. Co?), Nat., 717[1]; again, Riot, Y., p. 17.

Sensual Appetite, as he again finds man: „For yonder, lo, — — — see where the mad fool doth lie", F. El., p. 42.

Hypocrisy, as he sees the devil: „Sancti amen, who have we there?" L. J., p. 63, also

Nichol Newfangle: „Sancte benedicite, whom have we here?" L. W. L., p. 310.

Infidelity says the same, as Christ enters, M. M.[2].

Haphazard announces: „Judge Appius is come", A.V., p.131.

Ambidexter announces the approach of Sisamnes, K. C., p. 187, of Smirdis, p. 210.

Infidelity announces, „Yonder cometh Mary", M. M.[2].

Inclination (aside): „What, old doting Sapience here". T. T., p. 277.

Sedition observes Dissimulation who sings the litany as he enters: „I trow here cometh some hogherd calling for his pigs", K. J., p. 25.

Iniquity notices the approach of his accomplices: „I marvel who they be I see coming here". K. D., 175, again, as Equity departs: „Lo! he is gone". 524.

The Vice to Horestes: „. . . but harke, at hand Egistus draweth nye" etc., O., 753.

Ambidexter calls the king's attention to the arrival of the musicians. H. C., p. 236.

5. *The Vice reports what occurs behind the scenes.* — [Man., Nat., K. C., A. V., Confl., O., T. T., Tide].

Mischief:
„He (Mercy) hath taught Mankind, wyll I haue been
[frame,
To fyght manly ageyne hys fone". Man., 404—5.

Sensuality reports the fisticuff between man and Reason, including his own participation therein, Nat., — 1169[I], also the employment of the man whom he had led astray:

„He ys besy, harke in your ere, with lytell Margery",
[etc., Nat. 338[II];

Inclination likewise:
„Cogitation and he in one bed doth lie". T. T., p. 277.

Hypocrisy:

„Such chopping cheer as we have made — — —
And who so pleasant with my lord as is Philologus?"
[Confl., p. 115.

Haphazard:

„Claudius is knocking with hammer and stone,
At Virginius' gate as hard as he can lay on", A. V., p. 134.

The Vice: „The marriage celebrated at the church I did see", O., 1074.

Ambidexter brings this motif to a high state of development. In many monologues he reports the actions, which have just taken place, but which have not been represented on the stage. First, the mourning at court, for Smirdis, who has been killed by the order of the king; the murder itself took place on the stage: „O, the passion of God, yonder is a heavy court: Some weeps, some wails, and some make great sport". K. C., p. 217. Then he too weeps and makes remarks about the king: „But hath not he wrought a most wicked deed?" p. 218. Secondly, the festivities at court upon the occasion of the king's marriage:

„O, the passion of me! marry, as ye say, yonder is a
[royal court,
There is triumphing, and sport upon sport, ⨯ ⨯ ⨯
Running at tilt, justing, with running at the ring,
Masking and mumming, etc. ⨯ ⨯ ⨯
Such dancing and singing etc., ⨯ ⨯ ⨯
O, there was a banquet royal", etc., p. 231, 232.

With this he unites remarks about the expense of banquet and also about marriage in general.

Thirdly, about the mourning for the queen, whom the king has caused to be killed:

„Ah, ah, ah, ah, I cannot choose but weep for the queen,
Nothing is worn now but only black" etc., p. 243.

He also weeps, but only in burlesque, for he says:

„O, o, my heart, my heart: o, my bum will break".

In this monologue he recounts all the cruelties of the king:

„Cambises put a judge to death; that was a good deed;
But to kill the young child was worse to proceed;

To murder his brother, and then his own wife!
So help me God and halidom, it is a pity of his life". p. 244.
Courage enters weeping:
„Out alas, this tidings are ill,
My friend Greediness hath ended his days,
Despair upon him hath wrought his will,
Yes, truly he died and went with the tide-boat straight
[into hell,
And some said that he died of the new sickness", Tide, Gii.

6. *The Vice makes side-remarks about his fellow-players; these remarks are generally sarcastic.* — [W. C., Confl., M. W., L. W. L., M. M.²].

Folly after getting Manhood wholly in his power, says:
„Ah, ah, sirs, let the cat wink, ×××
I shall draw him such a draught of drink ×××
Lo, sirs, thus fareth the world away". W. C., p. 265.

Hypocrisy says, as Philologus yields to the temptation:
„We have caught him as a bird in lime". Confl., p. 99.

Iniquity says, as Ismael is brought in as a bound captive:
„Ye be tied fair enough for running away", N. W., p. 176.

Similarly, Nichol Newfangle says as Cuthbert, Cutpurse and Pierce Pickpurse are being led away:
„Ha, ha, ha! there is a brace of hounds — — —
Behold the huntsmans leadeth away". L. W. L., 355.

Infidelity exhibits an ugly disposition; as Mary Magdalene yields to his persuasion and thanks him for his advice, he comments thus: „Verba puellarum folliis leniora caducis". M. M.². These words he translates and explains.

———

Resume. The **chief facts** shown in the foregoing pages are the following:

I. Of the devil.

    1. In the non-dramatic literature the devil-scenes, excepting in the Legends, are restricted to certain biblical precedents.

    2. The same is largely true of the Mystery-cycles.

3. In the Digby Plays and Noah's Ark the figures of the devils are the same as those of the Legends.

4. The character of the devil on the stage has not been developed in a popular sense; he is comical or satirical only to a limited extent.

5. The devil ceased to be an important person on the stage as early as 1500.

II. Of the Vice.

1. The figure of the Vice is not derived from that of the devil but rather from the Seven Deadly Sins.

2. The character of the Vice is three-fold:
   - α) As an enemy of the Good and as a satirist.
   - β) As a tempter of man.
   - γ) As a buffoon.

3. The Vice is distinct from the clown and the fool.

4. The Vice disappeared from the stage with the disappearance of the Moralities.

5. The figure of the Vice has been introduced into the Tragedies only to a very limited extent.

6. In the latter part of the sixteenth century the name „the Vice" came to be applied to the buffoon simply.

Bibliography.

Arnold, Thomas, Wiclif's select English works, 1871.
Bale, John, King John, Camden Soc., ed. Collier, 1838.
Böddeker, K., Altengl. Dichtungen des Harl. M. S., 1878.
Bokenam, Osbern, Legenden, ed. Horstmann, 1883.
Bolte, Johannes, Der Teufel und die Kirche, Zeitschrift für vergleichende
    Litteratur, 1897, p. 249.
Brandl, A., Quellen des weltlichen Dramas in England vor Shakespeare,
    1898.
—, All for Money (Manuscript copy in the possession of Professor Brandl).
—, The Tide tarrieth for no man, (Manuscript copy in the possession of
    Professor Brandl).
—, Maria Magdalene (Manuscript copy in the possession of Professor
    Brandl).
Brotaneck, Rud., Das Noahspiel, Anglia XXI.
Chaucer, Geoffrey, The Students', ed Skeat, 1897.
Collier, J. Payne, The History of English Dramatic Poetry, 1831.
—, John Bale's King John, Cam. Soc., 1838.
—, Albion Knight, Shakespeare Soc., 1844.
—, Performance of Dramas by the Parish Clerks and Players in Churches,
    Shakespeare Soc., Papers, 1847.
Cotgrave, Randle, A Dictionary of the French and English Tongues,
    1611.
Creizenach, Wilhelm, Geschichte des neueren Dramas, Halle, 1893.
Dodsley, Old English Plays, ed. Hazlitt, 1874.
Douce, Francis, Illustrations of Shakespeare, 1807.
Dryer, Max, Der Teufel in der deutschen Dichtung des Mittelalters.
    Rostock, Diss., 1884.
Dyce, Alex., Skelton's Poetical Works, 1843.
—, Sir Thomas More, A Play, Shakespeare Soc., 1844.
—, The Poetical and Dramatic Works of Robert Greene and George
    Peele 1889.
Eberling, F. W., see Flögel.
Einenkel, E., The Life of Saint Catharine, Early English Text Soc.,
    1884.

Fairholt, F. W., John Heywood, Percy Soc., XIX—XX, 1842.
—, Costume in England, 1860.
Fischer, Rudolf, Zur Kunstentwickelung der englischen Tragödie, 1893.
Flögel, Geschichte des Grotesk-Komischen, neubearbeitet von Eberling, Leipzig, 1862.
—, Hofnarren.
Furnivall, F. J., The Digby Mysteries, The New Shakespeare Soc., 1896.
Gaspary, Adolf, Geschichte der italienischen Litteratur, 1885.
Genée, Rudolph, Die Miracelspiele und Moralitäten, als Vorläufer des englischen Dramas, 1878.
Gifford, William, The works of Ben Jonson, 1816.
Graf, Herman, Der miles gloriosus im engl. Drama. Diss., Rostock, 1891.
Green, H., Witney's choice of Emblems, 1866.
Halliwell, J. O., The Coventry Mysteries, Sh.-Soc., 1841.
—, Tarlton's Jests, Shakespeare Soc., 1844.
—, A Dictionary of archaic words etc., 1847.
—, A Dictionary of English Plays.
—, The marriage of Wit and Wisdom, an ancient Interlude. Sh.-Soc., 1846.
—, The moral Play of Wit and Science, Shakespeare Soc., 1848.
Hase, Karl, Das geistliche Spiel, 1858.
Hazlitt, Wᵐ, A View of the English stage, 1818, Ed. Dodsley.
Herford, C. H., Studies in the Literary Relations of England and Germany, 1886.
Holthausen, F., Das Noahspiel, Göteborg, 1897.
Hone, William, Ancient Mysteries, 1823.
Horstmann, C., The Early South Engl. Legendary, Early Engl. Text Soc., 1887.
—, Lives of the women Saints of England. Early Engl. Text Soc., 1886.
—, Altenglische Legenden-Sammlung, 1878.
—, Osbern Bokenam's Legenden, 1883.
Hünerhoff, August, Ueber die komischen „Villain" — Figuren der altfranzösischen chansons de geste, Marburg, Diss., 1894.
Jamieson, T. H., Barclay's Ship of Fools, Edinburgh, 1874.
Johnson, Samuel, Dictionary, 1755.
Jonson, Ben, Works, see Gifford.
Klein, J. L., Geschichte des Dramas, 1886.
Langland, William, Piers the Plowman, ed. Skeat, 1881.
Lyndesay, Sir David, Works, Early Engl. Text Soc.
Mall, E., The Harrowing of Hell, Breslau, 1871.
Malone, Edmond, The Works of Shakespeare, 1790.
Marlowe, Christopher, Doctor Faustus, ed. Breymann, 1889.
Morris, R., Genesis and Exodus, E. E. T. S., 1865.
—, The Blickling Homilies, E. E. T. S., 1874.
—, Old English Homilies, E. E. T. S., 1867—73.
—, Cursor Mundi, E. E. T. S., 1874.
Osborn, Max, Die Teufellitteratur des XVI. Jahrh. Acta Germanica III, 1893.

148

Pollard, A. W., English Miracle Plays, 1895.
—, The Townley Plays, E. E. T. S., 1897.
Puttenham, George, The Arte of English Poesie, 1589. Arber, 1869.
Reuling, C., Die komische Figur in den deutschen Dramen bis XVII. Jh. Zürich, Diss., 1890.
Roskoff, Georg Gustav, Geschichte des Teufels, Leipzig, 1869.
Schneeganz, Geschichte der grotesken Satyre, Strassburg, 1894.
Schröder, Karl, Das Redentiner Spiel, 1893.
Shakespeare, W., Works, Globe Edition.
Sharp, James, Dissertation on the Coventry Mysteries, 1825.
Skelton, John, Poetic Works, ed. Dyce.
Skeat, W. W., Aelfrics Lives of the Saints. E. E. T. S., 1881.
—, The Students' Chaucer, 1897.
—, The Vision of Piers the Plowman, 1881.
Smith, Lucy T., The York Mystery Plays, 1885.
Strutt, Jos., View of the manners and customs of the Inhabitants of England, 1775.
Swoboda, W., John Heywood als Dramatiker, Wien, 1888.
Symonds, J. A., Shakespeare's Predecessors, 1884.
Ten Brink, Bernhard, Geschichte der englischen Litteratur II, 1893.
Theobald, Lewis, The Works of Shakespeare, 1767.
Ticknor, George, History of Spanish Literature, 1849.
Thümmel, J., Die Charaktere Shakespeares, Halle, 1887.
Triggs, Oscar L., Lydgate's Assembly of the Gods. E. E. T. S., 1896.
Wager, Lewis, Maria Magdalene (see Brandl).
Walpul, George, The Tide tarrieth for no man (see Brandl).
Ward, A. W., A History of English Dramatic Literature, 1875.
Weinhold, K., Ueber das Komische im altdeutschen Drama. Jhb. für Geschichte und Litteratur, 1865.
Wessely, J. E., Die Gestalt des Todes und des Teufels, 1876.
Wieck, H., Die Teufel auf der mittelalterlichen Mysterienbühne Frankreichs, Marburg, Diss., 1887.
Wright, T., The Chester Plays, Shakespeare Soc., 1843—1847.
—, A History of Carricature and Grotesque, 1865.
Wurt, L., Das Wortspiel bei Shakespeare, Wien, 1895.
Wyclif, John, Select Works, ed. by Arnold, 1871.

Printed and bound by CPI Group (UK) Ltd, Croydon, CR0 4YY

23/10/2024

01778242-0002